Irrigation and River Basin Management

Options for Governance and Institutions

Irrigation and River Basin Management

Options for Governance and Institutions

_____

Edited by

Mark Svendsen

Consultant, Oregon, USA
and
Fellow, International Water Management Institute

CABI Publishing
in association with the
International Water Management Institute

CABI Publishing is a division of CAB International

CABI Publishing
CAB International
Wallingford
Oxon OX10 8DE
UK

Tel: +44 (0)1491 832111
Fax: +44 (0)1491 833508
E-mail: cabi@cabi.org
Web site: www.cabi-publishing.org

CABI Publishing
875 Massachusetts Avenue
7th Floor
Cambridge, MA 02139
USA

Tel: +1 617 395 4056
Fax: +1 617 354 6875
E-mail: cabi-nao@cabi.org

Published in association with:
International Water Management Institute
PO Box 2075
Colombo
Sri Lanka
www.iwmi.cgiar.org

A catalogue record for this book is available from the British Library, London, UK.

Library of Congress Cataloging-in-Publication Data
Irrigation and river basin management : options for governance and institutions / edited by Mark Svendsen.
   p. cm.
  Includes bibliographical references and index.
  ISBN 0-85199-672-8 (alk. paper)
  1. Watershed management.   2. Watershed management--Case studies.
  3. Irrigation.   4. Arid regions.   5. Integrated water development.
  I. Svendsen, Mark, 1945– II. Title.

TC405.I77 2004
333.91′3--dc22   2004007807

ISBN 0 85199 672 8

Typeset by AMA DataSet Ltd, UK.
Printed and bound in the UK by Biddles Ltd, King's Lynn.

Contents

———————————

v

Contributors

Charles Abernethy is a consultant on irrigation and water resources management, based in Colombo, Sri Lanka, where he was Senior Technical Advisor at the International Irrigation Management Institute until 1994. He works mainly in Asian countries, as a resource person assisting training courses and workshops on institutional and management questions related to water. E-mail: abernethy@itmin.com

Necdet Alpaslan is a civil engineer and specializes in water quality, wastewater treatment and management of solid wastes. He is a professor at Dokuz Eylul University, Izmir, Turkey, and the Director of Center for Environmental Research and Development (CEVMER). He has been working on various research and application projects, and giving consultancy services in these fields for 25 years. E-mail: nalpaslan@deu.edu.tr

Martin Burton is a Director of ITAD~Water Ltd, and an IWMI Fellow. He specializes in water management and capacity building, and is applying his recent MBA training to the management of change within the irrigation and water resources sector. He has lived and worked in over 25 countries. E-mail: martin.burton@itad.com

Nilgün Harmancıoğlu is a professor of hydrology and water resources, affiliated with Dokuz Eylul University (DEU) Faculty of Engineering in Izmir, Turkey, since 1976. Her areas of research are water resources management, watershed modelling, and decision support systems as applied to water resources. She is currently the director of the Water Resources Management Research & Application Center (SUMER) of DEU and head of the Water Resources Department of the University. E-mail: nilgun.harmancioglu@deu.edu.tr

Marna de Lange is a South African who trained as an engineer and works in rural development and poverty alleviation, with a specific focus on household food security. Her activities span hands-on implementation, training and capacity building of development workers, research and policy brokering. E-mail: marna@global.co.za

Hervé Lévite is a Researcher based at IWMI's Africa office in Pretoria. He is an engineer by training and works on water resources management issues under the co-supervision of CEMAGREF (France). E-mail: h.levite@cgiar.org

Ian Makin is IWMI Regional Director in Southeast Asia. He was trained as an irrigation engineer and has extensive research experience in water resources and irrigation system operations and management in South and Southeast Asia developed over 27 years working in many countries of the region and in Africa. E-mail: i.makin@cgiar.org

Douglas J. Merrey is a social anthropologist who has worked on water and irrigation institutional issues since the mid-1970s. He is presently IWMI's Director for Africa. His research is focused on the institutional arrangements at local, river basin and national levels for management of water resources, especially for irrigation. He has lived and worked in Pakistan, India, Sri Lanka, Indonesia, Egypt and South Africa, and has done shorter-term assignments in a number of other developing countries. E-mail: d.merrey@cgiar.org

David Molden is a Principal Researcher at the International Water Management Institute based in Sri Lanka. He has 25 years experience in water resources in the fields of irrigation, groundwater, river basin analysis and institutions, with resident assignments in Botswana, Egypt, Lesotho, Nepal and Sri Lanka. E-mail: d.molden@cgiar.org

François Molle is a Senior Researcher at the Institut de Recherche pour le Développement in France, and currently serves at IWMI in Colombo. His main areas of work and points of interest are irrigation and river basin management and water policy. E-mail: f.molle@cgiar.org

Hammond Murray-Rust was team leader for IWMI in Turkey, and subsequently Theme Leader for IWMI's research on integrated water management for agriculture based in Colombo, Sri Lanka, and then Hyderabad, India. He is now Senior Associate at ARD Inc in Burlington, Vermont, USA. He specializes in integrated modelling for water management, determinants of productivity of water, irrigation performance assessment, and use of remote sensing for water management. E-mail: hmurray-rust@ardinc.com

Claudia Ringler is a Research Fellow with the International Food Policy Research Institute in Washington, DC. Her research interests are in river basin management and natural resource policies for developing countries. E-mail: c.ringler@cgiar.org

Ramaswamy Sakthivadivel is an IWMI Senior Fellow stationed at Chennai, India, and works half time with IWMI as consultant. Trained as a civil engineer, he specializes in irrigation and water management with special reference to groundwater and small-scale irrigation systems. He has been with IWMI for more than 15 years, working in a dozen countries on his area of specialization. E-mail: sakthivadivelr@yahoo.com

Madar Samad is Theme Leader for Water Resources Institutions and Policy at the International Water Management Institute, Colombo, Sri Lanka. He is an agricultural economist by training. His research interests are irrigation and water resources institutions, and agricultural development. He has carried out research in these fields in several Asian and African countries during the last 10 years. E-mail: m.samad@cgiar.org

Christopher Scott is a hydrologist with over 15 years of research and applied hydrology and water resources expertise, including 6 years research and management experience with IWMI in Mexico, USA and India. He is currently IWMI South Asia Regional Director, based in Hyderabad, India. E-mail: c.scott@cgiar.org

Tushaar Shah is a Principal Scientist at IWMI based in Gujarat, India. He leads IWMI's global research on sustainable groundwater management. Trained as an economist and management specialist, Shah's work has focused on irrigation and water management institutions in Asia and Africa. E-mail: t.shah@cgiar.org

Nguyen Duy Son is currently an Operations Officer with the World Bank in Hanoi. Formerly, he was a Land and Water Resources Planner with the Sub-Institute for Water Resources Planning in Ho Chi Minh City, Vietnam. Trained as an engineer, he specializes in integrated land and water resources development and management, and has worked on these issues throughout Vietnam over the course of 10 years. E-mail: ng-duyson@hcm.vnn.vn

Mark Svendsen is a consultant based in Oregon, USA, and an IWMI Fellow. Trained as an engineer, he specializes in irrigation and water resource institutions, and has worked with these issues for a wide variety of clients in 25 countries in Asia, Africa, Europe, and North and South America over the course of 30 years. Prior to becoming an independent consultant, he was a Research Fellow at the International Food Policy Research Institute in Washington, DC, USA. E-mail: marksvendsen@aol.com

Henri Tardieu is General Manager of the Compagnie d'Aménagement des Coteaux de Gascogne, a semi-public irrigation and water resources service provider in southwestern France, and a member of the board of a French basin agency. As an engineer, he is also an international consultant in the field of water management and institutional evolution. E-mail: h.tardieu@cacg.fr

Philippus Wester is Assistant Professor in the Water Reforms, Irrigation and Water Engineering Group at Wageningen University in the Netherlands. Trained as a socio-technical irrigation engineer, he has worked as a water management researcher in Senegal, Pakistan, the Netherlands, Bangladesh and Mexico in the past 10 years, focusing on irrigation management transfer, participatory water management and river basin management. His current research interests centre on the production and articulation of water reforms and their effects. E-mail: flip.wester@wur.nl

Preface

When ideas and discussions chain together, with one idea leading to studies and then additional discussion and more new ideas, it is difficult to pinpoint the origins of the discussion thread. That is the case with this book. In the 'post-IMT' world of the late 1990s, questions arose about relationships between newly independent Water User Associations (WUAs) and the national water agencies that spawned them. While the WUAs needed to gain experience with independent operation, national water agencies recognized that they must continue to support the new WUAs with training, advice and other assistance if they were to make the transition to independent operation successfully and permanently. This led to IWMI research on support services required by fledgling WUAs. Of course, this line of inquiry itself grew directly out of the experience earlier in the decade with the challenges of establishing and empowering the WUAs in the first place.

In addition, concern arose over the relationship between the irrigation commands managed by the new WUAs and the hydrology of the basins in which they were situated. More specifically, questions concerned the impact that water scarcity and competition in the basin would have on WUAs operating within the basin. No longer wards of the state, irrigation systems, and their WUA managers, were now more vulnerable to the competitive pressures affecting water use in the basin. In addition to managing their own internal affairs, WUAs also had to be concerned with the security of their water supply and the impacts of the actions of other water users in the basin on its quality and availability.

These concerns led, in 1999, to discussions among Doug Merrey, Tushaar Shah, Martin Burton, Hammond Murray-Rust, David Molden and myself about the possibility of developing a methodology to assess basins simultaneously in terms of their hydrology and their institutional make-up. The feeling was that there were important connections between the two that could be elaborated. At the time, we called this exercise 'hydro-institutional mapping'. This led to the commissioning of several case studies on river basin management and to several conceptual papers such as Tushaar Shah's stimulating 'Limits to Leapfrogging'. It also led to a matrix template, which allowed the mapping of *essential basin functions* on to *basin actors* in order to examine, in a structure way, who was doing what in a basin.

IWMI research on this whole set of issues had been funded by the German government for several years, and this work culminated in a workshop on 'Integrated Water Management in Water-stressed River Basins in Developing Countries', held at Loskop Dam in South Africa in October of 2000. This workshop focused primarily on river basin management institutions and organizations under conditions of water scarcity, but the concern with the implications of scarcity for WUAs was always in the background. This book is an outgrowth of that workshop.

Seven of the 13 chapters in this book began as papers presented to the workshop. They comprise three of the book's thematic papers and four of the case studies. Most were extensively modified following the workshop before being included in the book. To supplement the three thematic chapters, I worked with Flip Wester and François Molle to write a conceptual paper on river basin institutions to complement the Molden *et al.* chapter on basin closure. I also asked Martin Burton to develop a chapter on the role of information in basin management, information being nearly as important a resource as water itself in managing a basin. To fill out the set of case studies, Marna de Lange *et al.* prepared a study on South Africa's experience with establishing a representative Catchment Management Agency in the Olifants Basin, and I contributed a case study on Vietnam's Dong Nai River Basin to add a Southeast Asian dimension. The Vietnam case was originally developed for the International Food Policy Research Institute with support from the Asian Development Bank.

Because the earlier discussions on 'hydro-institutional mapping' had involved many of the case study authors, there was a good deal of consistency in the way the authors approached the case studies, which was enhanced further during the interactive editing process. As one result, all of the studies employed the 'essential functions' matrix as a heuristic device and all focused, at least in part, on critically important basin governance issues. This lent a coherence to the cases that facilitated a cross-cutting look at them, a look which took the form of a pair of analyses – one focusing on basin management itself and the other looking at implications of basin closure for small-scale irrigators. These analyses comprise the final two chapters in the book.

Clearly there are important connections between water scarcity and the institutions that evolve to manage water-scarce basins, but while scarcity may drive the institutions' creation, it is local human, institutional and financial resources, economic forces, political dynamics, and, sometimes, international experience which shape them. The thematic chapters in the first part of the book develop the concepts used in the remainder of the book, particularly the ideas relating to closing basins and institutional governance, and offer suggestions, insights and qualifications related to those concepts. The case studies then describe some of the sharply differing forms that basin management institutions have taken in response to the influences mentioned above. The two analytic chapters attempt to understand what drives the creation of basin management institutions, describe how they differ from each other along key dimensions, and why, in some cases, they have taken the shape they have. We hope you, readers, will find these chapters interesting and useful.

I must first and foremost thank the 19 chapter authors, who of course made the book possible. Their biosketches are included in the list of contributors. Doug Merrey and Tushaar Shah were originally to edit this volume with me, but graciously turned the project over to me when I turned out to be the least busy of us. Our discussions during the book's formative moments were invaluable. Michael Devlin and Nimal Fernando of the IWMI publications office took my edited manuscripts and worked with the publisher to transform them into a book and, in addition, were very patient with me and my sometimes elastic timetable. Many other people were also involved in one way or another whom I will refrain from trying to list because I will inevitably forget someone important. Thank you all.

Mark Svendsen
Corvallis, Oregon
March 2004

1 Managing River Basins: an Institutional Perspective

Mark Svendsen, Philippus Wester and François Molle

1.1 Introduction

Up until the beginning of the 19th century, human water use was largely confined to streamside uses for drinking, stock watering and water-powered mills, as well as in-stream use for navigation. There were exceptions in the ancient hydraulic civilizations in Mesopotamia, Egypt, China, India and a few other locations, which abstracted large volumes of water from major rivers to irrigate extensive tracts of riparian floodplain. However, whereas the local impacts of these abstractions were significant, on a global scale, river flow regimes were still largely dictated by natural features and forces, and water users were primarily natural biota.

As the world population grew, from less than 1 billion (10^9) in 1800, to 1.7 billion in 1900, to more than 6 billion today, human demands for water also expanded. Growing urban concentrations, often along rivers, led to significant abstractions of water from rivers for these settlements and to negative impacts on water quality. At the same time, the industrial revolution created new demands for water, and new technology and the growing demand for food gave rise to an expanding irrigated agriculture throughout the world.

The latter half of the 19th century saw great strides in the development of hydraulic technology for controlling major rivers and transporting water over long distances for irrigation and domestic purposes. Many of these developments took place in the Asian subcontinent, and engineers from the USA and other countries pilgrimaged to British India to learn this new technology (e.g. Wilson, 1891). The following century witnessed a great remaking of river systems across the world, as humans manipulated the natural hydrology to meet the domestic supply, sanitation, food, fibre and industrial needs of growing populations and rising standards of living. During much of the 20th century, expanding water supply was the easiest and least costly way of satisfying these demands, since water was relatively abundant and the harmful impacts on the environment were incremental, individually modest and at first little noticed.

From a situation of limited, low-impact and largely riparian uses of water, we have now reached a point where, in many parts of the world, cumulative uses of river resources have not just local but basin-wide and regional impacts. The result is that water resources in many river basins are fully or almost fully committed to a variety of purposes, both in-stream and remote; water quality is degraded; river-dependent ecosystems are threatened; and still-expanding demand is leading to intense competition and, at times, to strife. In response, there is growing interest in management systems that can bring together fragmented water uses, and water users, into an integrated planning, allocation and management framework. A common element of these approaches is that they do not just cover a single water use or an administrative jurisdiction, but deal with an entire river basin or sub-basin, such as the Colombia, the Indus or the Limpopo.

Integrated management frameworks promises a number of important benefits:

- Greater utility from a given amount of water through adjusted allocations;
- Reduced groundwater mining through conjunctive management of ground and surface water;
- More intensive reuse of water through planned sequencing of uses;
- Improved water quality through more comprehensive data collection, monitoring and enforcement;
- Incorporation of current social and environmental values into water allocation and management decision making;
- Inclusion of a wider range of basin stakeholders into decision making;
- Reduced conflict among users.

Despite this promise, and although highly fashionable of late in policy circles, integrated river basin management (RBM) is rather rare in practice. Reasons for this include the following:

- It requires genuine collaboration among administrative and sectoral units;
- It usually involves reductions in discretionary authority on the part of existing managing agencies;
- Managers who would gain influence over basin decision making may currently be bureaucratically and politically weaker than current managers;
- Its costs can be significant;
- It creates uncertainty for present resource users;
- It makes planning and decision making more complex.

Given these potentially inhibiting factors, it is not so surprising that there are not more practising examples of integrated management of water resources at the basin level.

In order to hurdle these constraints, dissatisfaction with the current situation must be intense, and the prospective benefits of a new management regime significant. It is these conditions that we explore in this chapter.

This chapter defines the basic elements and concepts comprising integrated basin management and other key concepts and then focuses on the process of analysing institutional arrangements for RBM to further our understanding of institutional design. To do this, we first discuss institutions, organizations and policies in relation to water management. We then outline an essential functions and enabling conditions framework for analysing basin management regimes and discuss possible institutional arrangements for RBM that meet the needs of locally managed irrigation.

1.2 Terms and Concepts

1.2.1 River basin water resource management

There are a number of terms used to describe an integrated process of assessing and managing water resources at the basin level. Most are variations on the terms *integrated water resource management* or *river basin management (RBM)*.

Integrated water resource management (IWRM) is a newer term and is defined by the Global Water Partnership (TAC, 2000) in the following way:

> IWRM is a process which promotes the co-ordinated development and management of water, land and related resources, in order to maximize the resultant economic and social welfare in an equitable manner without compromising the sustainability of vital ecosystems.

IWRM tends to have a strong normative content, often referring to the Dublin Principles and emphasizing such values as economic benefit, equity, sustainability and public participation. It is implicitly suggested that all these values can be made commensurate and compatible, but in practice, there are often trade-offs between them, particularly between equity and economic efficiency, and between economic welfare and sustainability.

RBM is a more traditional term and has more recently broadened its meaning to encompass many of the same features and values which characterize IWRM. RBM is defined by Mostert *et al.* (2000) as follows:

> RBM is the management of water systems as part of the broader natural environment and in relation to their socio-economic environment.

This definition is simpler and less overtly value-laden than the previous one. Both definitions encompass both planning and management of water resources, even though the planning element is implicit. A key element of both concepts is that planning and management units almost always cut across other divisions more traditionally used to manage resources, such as sectors, provinces or even nations, and herein lies both their great strength and their challenge. Some divisions across which the approach we are discussing extends are shown in Box 1.1.

Typically, the boundaries of the management unit are defined hydrologically, as a basin or a sub-basin, defined by Mostert *et al.* (2000) as follows:

> A river basin is the geographical area determined by the watershed limits of a system of waters, both ground and surface, flowing to a common terminus.

In a few areas, concepts do diverge somewhat. Basin management can readily be extended to management of other related resources in a basin, especially land resources. IWRM focuses more tightly on the water resource. At times RBM also extends into the realm of river basin development, in which case project-based development of new basin infrastructure often acquires a dominant position in the paradigm. Millington (2000) notes that project design or evaluation, construction and operation is still a very strong role of

---

Box 1.1. *Integrated* basin management.

Although the term *integrated* most commonly refers to integration across use sectors, such as agriculture and urban water supply, it can also encompass a number of other divisions, including the following:

- Administrative jurisdictions
- Ground and surface water
- Upstream and downstream reaches
- Environmental and human uses
- Supply and demand management
- Water quantity and quality
- Land and water use
- Trans-boundary uses

---

many river basin organizations. Experience has shown that where project-based design and construction is a major activity, resource management functions tend to receive low priority. IWRM grows out of a 'post-construction' milieu, where basins are closing (Chapter 2) and new construction tends not to be the dominant activity.

One important caveat to all this is that integrated basin management does not imply or require a single basin management organization. There is an unfortunate tendency in some quarters to equate basin management with a unitary basin management organization and to assume that in the absence of such an organization, effective integrated management is not possible. This is most certainly an incorrect assumption, as experience in the western USA demonstrates, and, in general, monolithic management organizations are the exception rather than the rule. The prototype for such thinking is probably the American Tennessee Valley Authority (TVA), which was, in its original form, an integrated basin development authority established during the economic depression of the 1930s to work in a very underdeveloped region of the country.[1] The TVA took on a wide range of development functions, including water resource development, in an area in which there was a dearth of effective organizational coverage by other public entities. The failure

---

[1]   The TVA has since evolved into an energy production and management agency.

of most other attempts patterned on the TVA, such as the Damodar Valley Authority in India in the 1950s, suggests the uniqueness of this model. More common among effective basin management set-ups is a coordinated model in which the efforts of a number of different entities are articulated (see Section 1.3.4).

1.2.2 Institutional arrangements

How water is used in a river basin is determined by the interactions between water users, technology and water availability, and hence is a sociotechnical process (Mollinga, 1998). Water management organizations and institutions structure and mediate these interactions, and in turn are reshaped by water use in practice.

The terms 'institutions' and 'organizations' are often used interchangeably, but it is useful to distinguish between them. In mainstream institutional theory, *institutions* are understood to be 'the humanly devised constraints that shape human interaction' (North, 1990, p. 3), and consist of complexes of norms, values and behaviours that persist over time and inform action (Uphoff, 1986). In this view, institutions provide structure and regularity to everyday life by reducing uncertainty and providing a guide to human interaction. They are what Sir V.S. Naipaul calls 'the contract between man and man'. A central tenet of this view is that institutions work to reduce transaction costs by reducing the costs of monitoring and responding to the behaviour of others.

Organizations, on the other hand, are defined as 'groups of individuals bound by some common purpose to achieve objectives' (North, 1990, p. 5). Other definitions highlight the importance of seeing organizations as 'structures of recognized and accepted roles' instead of only groups of individuals, yielding a more realistic and accurate definition (Uphoff, 1986). Organizations are created intentionally within an existing web of institutions. Hence, what types of organizations exist and how they evolve are fundamentally influenced by the

broader network of institutions in which they are embedded. Organizations, in turn, influence how institutions evolve over time. Organizations constitute a subset of institutions, which are distinguished by their purposive origin and maintenance and their hierarchically organized roles.

In this book, the combination of the institutions and organizations involved in water management is termed the *institutional arrangements* for water management. Institutional arrangements for water management thus include the following:

- The established policy and legal environment (policies, laws, rules, rights, regulations, conventions, and customs, both formal and informal);
- Water management organizations with responsibilities in water management;
- Processes, mechanisms and procedures for decision making, coordination, negotiation and planning.

At this point, a chasm opens, which must be carefully negotiated. Policies, rules and regulations, as specified by public authorities, may differ substantially from the application of those rules in practice. Moreover, local rules, such as those governing the allocation of water, for example, may be quite different from the formal set of rules promulgated by state authorities. The student of institutional arrangements must thus be aware of both the formal rules, and the set of *rules-in-use*, which operate on the ground. The importance of the differences that often exist between the two has led one prominent analyst to define institutions explicitly as rules-in-use, rather than simply as rules (Ostrom, 1992).

To this point, we have treated institutional arrangements as fixed and pre-existing. To animate this static view, it is necessary to consider how institutions emerge and how changes in them take place. Alternative approaches to the study of institutions, grounded in anthropology and sociology, argue that institutions are not only 'the rules of the game' or 'sets of working rules or rules-in-use' (cf. North, 1990; Ostrom, 1990) but are reproduced, transformed and subverted through interactions and negotiations between actors (Mosse, 1997; Cleaver, 1999; Mehta *et al.*, 1999). This approach suggests that institutions emerge historically from interactions, negotiations and contests between heterogeneous actors having diverse goals, and that they only continue to exist if they are invested in or practised. Thus, institutions cannot be seen apart from what people do, and are constantly made and remade through people's practices (Mehta *et al.*, 1999). Cleaver suggests that, '[t]he institutions for the management of water . . . are socially located and critically depend on the maintenance of a number of gray areas and ambiguity regarding rights of access, compliance and rules, [and] on a continuous process of negotiation between all users' (1999, p. 602). Such a notion of institutions opens avenues to analyse how power pervades institutional arrangements and gives rise to differentiated access to and control over water, and, more importantly, how to design processes to redress inequities.

Because water is essential to life and livelihood security and has multiple uses and users, water management readily gives rise to intractable or 'wicked' problems, especially where competition for water is acute. Wicked problems are clusters of interrelated problems characterized by high levels of uncertainty and a diversity of competing values and decision stakes (Ackoff, 1974; Rittel and Webber, 1973). The set of problems constituting a wicked problem cannot be solved in isolation from one another and are intractable since what constitutes a solution for one group of individuals entails the generation of a new problem for another. As wicked problems are characterized by competing perceptions and values, and often also involve power disparities, they enter the realm of politics, understood here broadly as the forum for choosing among values and the process through which relations of power are constituted, negotiated, reproduced or otherwise shaped (cf. Mollinga, 2001). Water is frequently a politically contested resource: a contest with unpredictable and unstable outcomes and diverging pathways to alternative futures (cf. Mosse, 1997; Mehta, 2000; Mollinga, 2001). Likewise, water management institutions and policies are frequently contested and the outcomes of political practices.

1.2.3 Institutional effectiveness

From the brief discussion of institutions and organizations above, it is apparent that studying effective institutional arrangements for water management is conceptually challenging. If institutions differ in principle and in practice, are contested, beset with ambiguities and the outcomes of political practices, it follows that what is defined as 'effective' by some will be deemed ineffective by others. Nonetheless, at an intuitive level, it is clear that the strong connection between institutions and how water is managed is indisputable. Heathcote (1998, p. 7) suggests that institutional arrangements for water management may be considered effective if they:

1. Allow an adequate supply of water that is sustainable over many years and provide equity in access to this water.
2. Maintain water quality at levels that meet government standards and other societal water quality objectives.
3. Allow sustained economic development over the short and long term.[2]

Thus, institutional arrangements for RBM are effective if they promote and achieve sustainable water management. Sustainability can broadly be defined as a condition in which natural and social systems survive and thrive together indefinitely (Euston and Gibson, 1995).

To be sustainable, water management must protect and restore natural systems, enhance the well-being of people and

---

[2] A fourth condition relating to maintaining ecological systems could be added here, but in our treatment is covered by an expanded definition of sustainability.

improve economic efficiency. These three objectives are often mutually exclusive, as the partial attainment of one has negative effects for attaining the others. This contested nature of water makes the institutional arrangements for water management of paramount importance, but also highly problematic. If institutions are viewed in managerial or interventionist terms, effective institutions are seen as those that contribute to attaining sustainable water management by reducing transaction costs, enhancing collective action and increasing certainty. If a more process-oriented and dynamic view of institutions is adopted, emphasis is placed on how institutions are embedded in power relations, and equity, and not economic efficiency, is a central concern.

1.2.4 Policy

Although established policies are included under our definition of *institutional arrangements*, they are often treated separately in discussions of 'policy and institutional arrangements'. One reason for this is that new policies not yet fully implemented do not necessarily rise to the level of 'institutions', as they have not demonstrated an ability to persist. Consequently, they are discussed separately here.

Policies provide a direction and suggest a course of action intended to influence decisions and actions in a particular realm of interest. Water resource policy thus gives overall guidance and direction to decisions and actions that determine the uses, protections and costs of water, and the subsidies and prohibitions related to its use. In the face of changing conditions, needs, priorities and values, policies and resulting actions must also change. Policies are important to water resource management because they can serve as important entry points into the established cycle of water management practices.

An instrumentalist view of government conceptualizes policy as a tool to regulate a population from the top down, through incentives and sanctions. Shore and Wright (1997, p. 5) summarize the conventional definition of policy as 'an intrinsically technical, rational, action-oriented instrument that decision makers use to solve problems and affect change'. Although many would agree that policies frequently fail to function as intended, there is a shared understanding that a good policy is one that adheres to the standard of rationality contained in the above definition.

However, conceiving of policy development and application as an unproblematic linear process that progresses from formulation to implementation to expected outcomes is dangerous, as it obscures how policies are produced through decidedly non-linear and non-rational means, through public and private negotiations, log-rolling,[3] political pressure, media manipulation, legal action and other processes involving a range of actors within and outside of government circles. Moreover, there is seldom a simple progression from policy formulation, to legislation, to framing regulations, to execution. The real process is much less tidy, with iterations, false starts and backtracking, where the lead role alternates between policy formulation and application.

An alternative perspective on analysing how policies lead to changes in water management practices conceives of policy formulation and implementation as political processes in which many interests are at stake. Premised on the notion that water management is a politically contested terrain (Mosse, 1997; Mollinga, 1998, 2001; Mehta *et al.*, 1999), the 'policy as process' approach attempts to understand how water policies are 'produced' by the interactions between water users, dominant paradigms and the institutional arrangements that mediate water control. Through these interactions, the content and composition of policies are redefined and transformed, frequently leading to very different results

---

[3]   A process in which legislators trade concessions and support for sometimes unrelated objectives.

from those envisioned. A politically informed analysis of policy processes helps to understand how policies work in practice to change control over water and water management, thereby giving insight into who gains and who loses.

1.2.5 Roles and actors

Individuals and organizations take action, and it is useful to define two other commonly used descriptive terms. *Roles* are sets of expectations and tasks associated with a particular function (Coward, 1980). As such, roles can be played by individuals or by organizations. *Actors* are those individuals or organizations who take actions in a particular context, and thus play roles. Actors can play a number of roles simultaneously, and roles can be split among different actors. Often actors, such as a government ministry, will play roles that relate to water resource management while playing other non-water-related roles at the same time. *Stakeholders* are individuals or groups which have a legitimate interest in the management of water resources in a basin but which may or may not play an active role in basin planning and management processes. Actors are thus included in the set of basin stakeholders, but do not comprise the entire set.

1.3. Basin Management

1.3.1 Context

1.3.1.1 Phases of basin closure

In Chapter 2, Molden *et al.* posit that river basins pass through three phases as more water is withdrawn by humans (development, utilization and reallocation), and argue the valid point that that institutions need to have the ability to adapt to changes. Keller *et al.* (1998) proposed a linear three-phase model of river basin maturation, with phases of exploitation, conservation and augmentation. In this model, the final

phase is a search for new water sources – from distant basins, or by desalinating seawater, rather than reallocation.

Turton and Ohlsson (2000), expand this general argument positing that water scarcity *per se* is not the key issue, but rather whether a society has the adaptive capacity to cope with the challenges that water scarcity poses. They argue that institutional transitions need to occur in the water sector as water becomes scarcer, the first when water abundance turns to water shortage and the second when water shortage turns to water over-exploitation. In their argument, the 'ability to cope' with shortage is a critical societal attribute determining the 'pain' a society will feel as a basin closes.

These models are inductive, attempting to draw out an explanatory thread from a body of experience. As such, they are heuristic rather than predictive in particular cases. Keller *et al.* (1998) emphasize the economic logic of the sequence of development. In this framework, at any point in time the cheapest solutions are selected, from simple flow diversion on to desalinization. Molden *et al.* constrain the impulse to continuously develop new sources of supply and instead suggest that attention will shift to reallocating an ultimately fixed supply. In Turton and Ohlsson's approach, the logic of the succession is based on a scale of complexity, with the solution of water scarcity problems demanding ever-increasing levels of social resources. This approach assumes that hydraulic development is the easiest response, and that its exhaustion leads to conservation efforts, later followed by adjustments in allocations. The latter are regarded as much more sensitive and likely to generate social conflicts. These three analytical grids are useful in making connections between the degree of water exploitation and types of human responses – responses that are clearly related to the degree of stress on the resource as well as other factors. At the same time, they cannot capture important nuances found in varied concrete situations. A number of interesting illustrations of this complexity are shown below (Molle, 2003).

- Sakthivadivel and Molden (2001) have compared five basins said to be at different stages of exploitation and have found that the problems faced by these basins were different. However, several of the problems encountered were not those that would be typical of the phase in which each basin was classified. For example, East Rapti basin in Nepal is an open basin, with only 5% of water resources being used by agriculture. In spite of this, water pollution from industries and 'intense competition' for river water during the dry season among wildlife sanctuaries, tourist requirements, ecological requirements and human use already appear as severe problems that 'need immediate action', although they are normally associated with later phases of development. In the Singkarak-Ombilin basin, Sumatra, considered to be at the beginning of the utilization/conservation phase, water allocated to non-agricultural activities and trans-basin diversion threatens to throw the basin directly into the last phase, where water rights need to be more formally specified and water reallocation becomes paramount.

- Problems of pollution are generally associated with late phases in which the scarcity of the resource does not allow adequate mitigation by dilution, but it may also happen very early if there are significant pollution point sources with little regulated water to ensure dilution as with gold mines in South Africa.

- The need to design more complex and integrated forms of organization at the basin level is associated with an ultimate phase of reallocation of very scarce water resources. However, in the case of France in the 1960s, it was the problem of water quality and not quantity that was the driving force, despite both aspects being interlinked.

- Trans-basin diversion is considered in trajectory models as a way to 'reopen' the basin after it has closed, but this option is sometimes observed at much earlier stages of development, especially in small and medium basins. In Sri Lanka, this was commonly achieved as early as the 5th century (Mendis, 1993) and has remained a basic feature of water resource development ever since. Such transfers are also typical of irrigation of mountainous interfluves, where irrigated areas straddle the boundary of two adjacent basins.

- In the later phases of basin closure, a wide range of measures are sometimes undertaken to relieve pressure, and not simply the reallocation strategy hypothesized. The case of California, as described by Turral (1998), clearly shows not only that both efficiency and reallocative measures are sought in parallel, but that the gains they provide are more limited than commonly believed and need to be accompanied with a substantial amount of supply augmentation. Closure does not conclude with reallocation but, rather, elicits continuous improvements on all three fronts (conservation, reallocation and supply).

- Not all trajectories are upward. Historical examples of civilizations that have not successfully maintained their resource base and have collapsed can easily be found. In Sri Lanka, for example, aerial photographs reveal a high density of abandoned, silted and destroyed small tanks in some basins. A classic case is that of ancient Mesopotamia in the 9th century (cf. Pointing, 1991).

It seems clear that decisions must be understood also in terms of their political economy. That is, decisions must be understood not only on the basis of their actual costs and social 'pain', but also in terms of the identity of the beneficiaries and the increased power or financial gain that accrues to different actors as a result of the decision taken. Costly solutions, such as desalination, are sometimes justified and implemented in lieu of less expensive demand-management solutions because they involve less political pain and can be accommodated under the existing logic of pork-barrel politics.

The difficulty of reforming management varies, depending on many cultural, social and political factors, but it is recognized that 'regional politicians have a powerful intuition that economic principles and the allocative measures which follow logically from them must be avoided at all costs' (Allan, 1999). This largely explains the persistent gap between consultant's rationality and the actual shape that policy measures take in the real world. It also suggests why strict economic analysis is not always the best framework to use to understand the succession of state investments and responses. Resource capture can occur at any time, depending on the power balance within the society, and is perhaps more frequent than rational allocation.

The basin closure models are useful in outlining hydrologic changes which tend to take place in a generalized basin as its water resources are utilized ever more intensively and the response strategies which tend to occupy the minds of basin managers as this unfolds. In practice, institutional responses are highly varied and functions of a diverse range of forces and influences.

1.3.1.2 The governance context

One of the most important, and least studied, aspects of the environment controlling institutional change is the context of governance in which water resource organizations operate. Governance is defined as the exercise of authority through formal and informal traditions and institutions for the common good. Governance includes: (i) the process by which those in authority are selected, monitored and replaced; (ii) the capacity of the government to effectively manage its resources and implement sound policies; and (iii) the respect of citizens and the State for the institutions that govern economic and social interactions among them (Kaufmann, 2000). In Kaufmann's framework, good governance consists of six interlinked components:

- Voice and accountability;
- Political stability;
- Government effectiveness;
- Lack of regulatory burden;
- Rule of law;
- Control of corruption.

Problems stemming from poor governance are numerous, well known and routinely given a blind eye. Examples include favouritism in granting water use permits, kickbacks on construction and procurement contracts, biased and inaccessible court systems, withholding of data and sale of public data for personal profit, promotion of risk-averse bureaucrats and firing of innovators, flaunting of water-quality regulations by well-connected industries, and bureaucratic red tape which strangles local initiative. Quality of governance pervades public decision making relating to policy formulation, resource allocation, legislation, rule enforcement and adjudication, making it the most important single influence on the shape and pace of institutional change in the water sector. While difficult to change, improving the quality of governance is not impossible, and a variety of tools have recently been developed to support such change, many involving voice, transparency, information and participation. Several specific elements, which can support improved governance and promote institutional change, are discussed in a subsequent section.

1.3.2 Functions and actors

To analyse the institutional arrangements for water management in a river basin we propose a framework of essential functions and enabling conditions. The groundwork for this framework is provided by drawing up a water account of a basin as well as a basin profile that provides an analytically rich description of the basin as a sociotechnical system. The next step is to identify the water management actors in a river basin and the essential functions they execute.

1.3.2.1 Actors

To portray the multitude of water management stakeholders in a river basin, it is useful to distinguish between the various water use sectors and the types of organizations involved in water resource management. Combining sectors with actors yields a matrix of key organizations and stakeholders involved in water management in a basin (Table 1.1). This matrix, after individual actors are identified under each category, provides a basis for identifying key actors in a particular basin.

1.3.2.2 Essential functions

To analyse basin governance, it is necessary to focus on the roles and functions of the various actors engaged in water management in the basin, asking who does what, where, to what end and how well. To guide the analysis, a set of essential functions for RBM has been identified (Burton, 1999; Svendsen *et al.*, 2001). These are defined in Table 1.2. How well functions are carried out, from whose perspective and for whose benefit are empirical questions.

It is possible to construct alternative lists of basin functions. This one was empirically and inductively developed from experience in a number of basins. Its functions subsume supporting functions such as data collection and resource mobilization, which are not ends in themselves, but rather facilitate the higher-level functions listed.

Table 1.1. Illustrative matrix of key water management stakeholders.

Stakeholders	Sector					
	Agriculture	Domestic	Industry	Hydropower	Environment	Other
Multinational agencies						
Government agencies						
Private firms						
Associations/NGOs						
Informal groups						

Note that *government agencies* may include national, sub-national and local entities, while *private firms* can include both multinationals and local firms as well as regulated for-profit utilities. The list of sectors is far from exhaustive and could also include fisheries, navigation, recreation, amenity value and others, depending on local importance.

Table 1.2. Essential functions for river basin management.

Function	Definition
1. Plan	The formulation of medium to long-term plans for the management and development of water resources in the basin, by which the water demands of different sectors are brought in line with water supply.
2. Allocate water	The mechanisms and criteria by which bulk water is apportioned among the different use sectors.
3. Distribute water	The activities executed to ensure that allocated water reaches its point of use.
4. Monitor water quality	The activities executed to monitor water pollution and salinity levels.
5. Enforce water quality	The activities executed to ensure that water pollution and salinity levels remain below accepted standards.
6. Protect against water disasters	Activities executed concerning flood and drought warning, prevention of floods, emergency works and drought preparedness.
7. Protect ecology	Actions undertaken to protect associated ecosystems.
8. Construct facilities	Activities executed for the design and construction of hydraulic infrastructure.
9. Maintain facilities	Activities executed to maintain the serviceability of the hydraulic infrastructure in the basin.

1.3.2.3 Interactional analysis

Functions that should be performed in a closing water basin for effective management may or may not be performed, in fact. Moreover, some may be performed incompletely and some may even receive more attention than they require. Since many organizations and stakeholders are involved with water management in a river basin, a number that generally grows with closure, more than one organization will frequently be involved in performing a particular function. To structure and clarify the resulting patterns of activity, the essential functions of basin management can be crossed with key water management actors as illustrated in Table 1.3. The essential functions are replicated, as appropriate, across three broad categories – surface water, groundwater and derivative water. Because the perspective is basin-wide, functions are not separated by sector, e.g. irrigation or environment. Thus, the category *derivative water* includes irrigation return flows as well as municipal wastewater and industrial discharges. Cells in the resulting matrix are coded to show whether the actor is judged to play a major or a minor role. One of the case studies in this volume (Chapter 9) adds another dimension to the table and endeavours to indicate the type of activity performed by an actor in addressing a particular function.

It is important to note that the matrix depicts actual activity in practice and not nominal responsibilities according to legal frameworks or normative prescriptions defining what should be done. The matrix, together with description of the key actors, gives an indication of which essential functions are being addressed and who is involved in their execution. The matrix exercise can shed light on a number of important questions:

- The functions covered and a rough indication of the adequacy of coverage;
- The functions not covered;
- The number of actors involved in each function and the need for coordination;

- The stakeholders represented in performing particular functions, which leads to conclusions about the representativeness of the basin governance.

Perhaps more importantly, the process is heuristic in that it produces insights and questions that can be used to probe more deeply into issues of functional performance, relationships among actors, and the political dynamics of basin governance and management.

The actors/functions matrix can also be used comparatively within a basin to examine changes over time, the nature of a transition to a desired future state, or nominal versus actual functional performance. Furthermore, it can be applied comparatively to look for patterns among different basins and national contexts. This is done in Chapter 13 of this volume for several of the case studies presented.

The matrix can be generated in different ways. It can be filled in by expert observers after study of a basin, as in the Turkey case study. It can be created on the basis of questionnaire survey results combined with expert observation as was done in the Mexico case, or it can result from focus group interactions of knowledgeable persons, as happened for the South African case. Although used here as a research tool, matrix generation can also be a useful understanding and consensus building tool among the involved parties when used in focus groups made up of key basin stakeholders.

1.3.3 Enabling conditions

Describing who executes the essential functions in a river basin and how effectively, while useful, does not constitute a sufficient methodology for understanding and diagnosing problems affecting basin governance. The essential functions and actors' roles depicted in Table 1.3 provide a static view of responsibilities. Additional attributes of well-functioning basin governance systems relate to its dynamics.

In order for societies to reach decisions consistent with the public interest, several

Table 1.3. Illustrative hypothetical matrix crossing essential basin management functions and key actors.

Key actors	Surface water									Groundwater							Derivative water					
	Plan (basin-level)	Allocate water	Distribute water	Construct facilities	Maintain facilities	Monitor quality	Ensure quality	Protect against disasters	Protect ecology	Plan (basin-level)	Allocate water	Withdraw/distribute water	Construct facilities	Maintain facilities	Monitor quality	Ensure quality	Plan (basin-level)	Allocate/distribute	Construct facilities	Operate/maintain facilities	Monitor quality	Enforce quality
Ministry of Environment	•	•					•	•	●						•	•					●	
Ministry of Water Resources	●	●		●	•	•		●														•
Ministry of Health	•					●	•		•						•		●	●			●	
River Basin Commission	●																●				•	
Electrical Authority	•	•						•		•												
Municipal Water Supply Company		•	●	●	●	●						●	●	●	●				●	●	●	
Irrigation System Office	•		●	●	●							●	●	●								
WUA			●	•	●					•	•											
Groundwater Department, Ministry of Water Resources	•														●							
Friends of the Environment (NGO)						●	•		•												●	
Leather tanning factory				•	•														•	•		
Association of Manufacturers (trade association)	●	●									●											
Federation of Irrigators (WUA umbrella group)	•	•									•											

●, Major role; •, minor role.
WUA, water user association; NGO, non-governmental organization.

conditions need to be satisfied. We term these attributes *enabling conditions* (Box 1.2). Enabling conditions are features of the institutional environment at the basin level that must be present, in some measure, to achieve good governance and management of the basin. These attributes are not specific to any one actor, but apply to all actors and their interactions and comprise necessary (but not sufficient) normative conditions for good basin management. Most of them contribute to good governance, as discussed in an earlier section. Some basic enabling conditions are shown in Box 1.2. A thorough analysis of these factors is well beyond the scope of this chapter, but a number of them are described briefly.

1.3.3.1 Political attributes

An important political attribute is the representation of interests. In most river basins, some water users will be well represented, while others will not, and in the arena of political give-and-take, those without representation become losers. Industrialists and commercial farmers, for example, typically have ample financial resources, are well organized and have ready access to political decision makers. Poorer irrigators, on the other hand, are likely to be less well organized and consequently will be weakly represented. Their interests are often rather fragile. Water users associations (WUAs)

Box 1.2. Enabling conditions.

Political attributes
  Representation of interests
  Balanced power
Informational attributes
  Process transparency
  Information availability
  Information accessibility
Legal authority
  Appropriate institutions
  Adequate powers
Resources
  Human
  Financial
  Institutional
  Infrastructural

---

are likely to be intermediate, particularly if they are connected to the local political establishment and collaborate informally, sharing information and coordinating activities. Many times WUAs would benefit by establishing more formal linkages among themselves to allow a single spokesperson to represent them collectively in discussions over basin water allocation, water quality standards, potential irrigation return flow restrictions, and so on.

A serious failure of representation will frequently exist for the environment. Experience has shown that strong non-governmental organizations (NGOs) rooted in civil society are essential components of a political system making socially responsible choices about environmental issues. NGOs can serve as advocates for environmental values and for unrepresented future generations. At the same time, fund raising requirements may lead such mass-based organizations to take extreme and uncompromising positions on issues that must then be moderated in the give-and-take of political debate and decision making.

Hence, fully as important as the existence of representational bodies is the need for a rough balance of political power and influence among various interests. When power is one-sided, issues are not aired adequately and decisions are also one-sided. A key to the evolution of a suitable and balanced governance regime is maturation of non-government organizations and associations based in civil society, which can advocate for particular interests, coupled with the informational attributes described below.

1.3.3.2 Informational attributes

An essential enabling condition is the presence in the public domain of accurate and up-to-date descriptive information on water-related issues in the basin. Another is open public transactions, related to policies, plans, regulations, violations and sanctions. The first of these stipulations require that information on basin water allocations, reservoir positions, groundwater elevations, water-quality conditions, available resources, and so on be a part of the public record. Information collected with public funds should be available to the general public at little or no charge in the interest of sound and democratic public decision making. This disclosure condition applies to intra- and inter-departmental information relationships, as well as to those with the general public. The second stipulation, transparency of public proceedings, is similarly essential to fair democratic processes. Rent-seeking behaviour requires darkness and privacy to thrive, and conducting regulatory processes in full view of the public and the press is an effective antidote to such practices.

1.3.3.3 Legal authority

Establishing appropriate organizations requires suitable legal authority. This authority includes the right to exist, the right to a legal personality and suitable electoral procedures to ensure representative leadership of the organization. A legal personality usually includes the right to handle money and keep a bank account, enter into contracts, access the legal system and represent the membership in dealings with governmental agencies.

In some cases, existing legislation has been adapted to allow new water-related organizations to be established. The formation of Irrigation Associations in Turkey took this route. In other cases such as Mexico, new legislation has been written at the outset to facilitate establishment of new organizations and relationships.

1.3.3.4 Resources

Clearly, all four types of resources listed in Box 1.2 are needed for effective implementation of basin management activities. A potential problem is scattering of human and financial resources among a number of organizations, where each lacks a critical mass to be effective. In a context of cooperation, it is not necessary that resources be consolidated under a single administrative structure for effective implementation.

However, cooperation and coordination must be effective if a decentralized strategy is to work.

1.3.4 Organizational configurations

The choice of a river basin as a unit of management is based on a certain hydrologic imperative, controlled by gravity and topography. However, establishing basin boundaries is by no means automatic. There are choices involved in subdividing large basins into management units, in grouping small basins, and in deciding which natural basins and sub-basins are to receive priority attention. Moreover, there are often differences between surface water and groundwater divides and basins, where choices have to be made, and where water is imported from neighbouring basins, controversies have arisen over whether to include the watershed of the transferred water in the basin definition. Making such choices has an important bearing on basin management, as different boundaries imply different decision makers and possibly different decision outcomes. However, whether defining boundaries is straightforward or contentious, the defined basin becomes a political unit as well as a hydrologic one and questions immediately arise as to who will make decisions, and how (cf. Wester and Warner, 2002).

Mostert *et al.* (2000) posit three different types of organizational configurations for basin management. One is an *authoritarian model*, in which management is organized on hydrologic boundaries and a single organization makes basin decisions. The second is a *coordinative model*, in which the basin as a hydrologic unit is recognized, but many functions remain in the hands of traditional governmental units, and work is coordinated among these units. The third is management by existing organizations without coordination. The third model is, in reality, not a model of basin management at all but rather the business-as-usual backdrop against which the two other models can be contrasted.

We distinguish two basic organizational patterns for basin governance.[4] The first is the centralized (unicentric) model, in which a single unified public organization is empowered to make decisions regarding management of the basin. This centralized organization is not necessarily 'authoritarian', but does centralize authority under a governance process that may be more or less democratic. The second is the decentralized (polycentric, coordinative) model, in which the actions of existing organizations, layers of government and initiatives are coordinated to cover an entire river basin or sub-basin. While new structures may be created, the bulk of routine work is done by existing organizations that are not specific to the basin. Although both models are characterized by separations among the three basic roles of management, regulation and service provision, the firewalls between them are typically stronger in the coordinative model where separate organizations are involved.

In the real world, RBM structures are usually hybrids, relying to some extent on existing government structures to provide policy and direction, and perhaps execute particular management functions, and basin-specific organizations to collect data, and make certain circumscribed decisions. We describe the two hypothetical models briefly to illustrate the two poles of the continuum.

1.3.4.1 Centralized (unicentric) RBMs

A strength of the centralized model is that its operational span of control coincides with the boundaries of the basin. This internalizes upstream/downstream and other conflicts, making them easier to deal with, and it concentrates the decision-making authority needed to resolve disagreements. Disadvantages of the centralized model include the following: (i) as the organization will generally deal only with water, water will be isolated from other relevant policy sectors such as agriculture, environment and the economy; (ii) establishing a strong unified central authority presents a more challenging political problem than securing agreement for a coordinative body. The challenge here will be even greater for international river basins; and (iii) governance of a centralized organization raises challenging questions of broad stakeholder representation and accountability.

The most prominent examples of authorities are those having the development of a river basin as a primary mandate. The classic example is the TVA, which was created during the economic depression of the 1930s to address problems of poverty and unemployment in a particular region of the USA. Other examples are the Rio São Francisco Development Agency in Brazil and the Mahaweli Development Authority in Sri Lanka. When their primary development tasks are finished, these authorities often try, with varying degrees of success, to assume a broader resource management role.

1.3.4.2 Decentralized (polycentric) RBMs

The decentralized model addresses some of the weaknesses of the centralized model but contains others. On the plus side, it provides for a strong political base for action, since coordination involves voluntary agreement among participating jurisdictions. The coordinative process also leads to a more responsive governance process. Intersectoral linkages remain intact, as coordination is among individual states, nations or other jurisdictions responsible for a range of policy sectors, and such a set-up provides a natural base for decentralization of responsibilities. On the other hand, decision making can be cumbersome, coordinating costs may be high and political changes in participating jurisdictions can upset agreements.

These two models represent polar extremes, and actual arrangements are often blends of the two. In the Murray–Darling

---

[4] We focus here on organizations and governance and not on the way in which particular functions are executed. There are a range of options within each governance structure for providing services and executing other functions.

basin, for example, a cooperative Ministerial Council, made up of representatives of the four involved states and the federal government, sets policy while under it an authority-like Commission supports the Council and executes its decisions. A similar set-up exists in France, where a River Commission made up of local and national government representatives and users sets water policy, which is implemented by an associated Water Agency. Publicly held companies manage the distribution infrastructure and make bulk water deliveries to user associations. In the USA, formal bodies for managing river basins are rare, allowing some exceptions such as the TVA and the Delaware Basin Commission. Policy-making authority is distributed among a variety of federal and state agencies and departments. Coordination is achieved through a plethora of committees and working groups linking stakeholders into discussion and decision-making forums. Legislation and negotiated legally binding agreements are important instruments for establishing policy and practices, and the court system is routinely invoked to resolve disagreements and disputes. In California, a state water plan, updated every 5 years, provides a rolling framework for managing the state's water resources.

1.4. Basin Management and Irrigation

As basins close, irrigation systems within the basin are confronted with both internal and external challenges. Internal challenges require them to do more with less water, whereas the external ones require them to organize and act effectively to protect their interests. Dealing with both at the same time is difficult, and systems which address the internal challenges successfully before having to tackle the external ones will generally be better off.

Closing basins, by definition, are becoming water scarce, and newer rapidly growing sectors, typically urban and industrial users and the environment, will usually demand that irrigated agriculture, as the largest traditional user, use less water to free more of it for their growing needs. This puts pressure on irrigators to use water more efficiently, and may lead to retirement of less productive irrigated lands and transfer of their water rights to other users, as is presently happening in California's Central Valley. The cost of water will generally increase to reflect its growing scarcity, leading to pressure to grow higher-value crops to cover these costs. More efficient use of water requires that irrigation systems acquire new measurement and control technology, and more professional management.

These same pressures may emerge as challenges to agriculture's right to use basin water resources at all in legislative and legal arenas. Often agriculture began using water at a time when rights were not formally specified, which can make them less secure than the formalized rights allocated to industries and larger corporate irrigators which came later. Where water rights are merely implicit in the allocation priorities of a large public irrigation agency, risks also arise. Regulation, service provision and other functions of unitary public water agencies tend to be split up among several new agencies or departments as basins mature, giving basin managers a broader constituency and weakening their ability to defend their former clients. Basin-level decision-making forums will tend to include more actors and cover a broader range of issues, requiring that irrigators mobilize to represent their own interests vigorously and become conversant with a broader range of water-related considerations.

Irrigators located downstream of major population centres and industries will also need to seek protection for the quality of the water they receive. Urban concentrations with inland locations, such as Cairo or New Delhi, degrade significantly the quality of water reaching downstream irrigators, with impacts on human health and contamination of produce and soil. Individually, irrigators and small systems will have little or no ability to apply pressure for reduced pollutant loadings in their water supplies. Organized into a larger network, they can have influence.

References

Ackoff, R.L. (1974) *Redesigning the Future: a Systems Approach to Societal Problems.* John Wiley and Sons, New York.

Allan, J.A. (1999) Productive efficiency and allocative efficiency: why better water management may not solve the problem. *Agricultural Water Management* 40, 71–75.

Burton, M. (1999) Note on proposed framework and activities. Prepared for the IWMI/DSI/CEVMER Research Programme on Institutional Support Systems for Sustainable Management of Irrigation in Water-Short Basins, Izmir.

Cleaver, F. (1999) Paradoxes of participation: questioning participatory approaches to development. *Journal of International Development* 11, 597–612.

Coward Jr, E.W. (1980) Irrigation development: institutional and organizational issues. In: Coward, W.W. Jr (ed.) *Irrigation and Agricultural Development in Asia.* Cornell University Press, Ithaca, New York, pp. 15–27.

Euston, S.R. and Gibson, W.E. (1995) The ethic of sustainability. *Earth Ethics* 6, 5–7.

Heathcote, I.W. (1998) *Integrated Watershed Management. Principles and Practice.* John Wiley and Sons, New York.

Kaufmann, D. (2000) Governance and anti-corruption. In: Thomas, V. (ed.) *The Quality of Growth.* Oxford University Press, Oxford, pp. 135–168.

Keller, J., Keller, A. and Davids, G. (1998) River basin development phases and implications of closure. *Journal of Applied Irrigation Science* 33, 145–163.

Mehta, L. (2000) Water for the twenty-first century: challenges and misconceptions. IDS Working Paper 111. IDS, Brighton.

Mehta, L., Leach, M., Newell, P., Scoones, I., Sivaramakrishnan, K. and Way, S. (1999) Exploring understandings of institutions and uncertainty: new directions in natural resources management. IDS Discussion Paper 372. IDS, Brighton.

Mendis, D.L.O. (1993) Irrigation systems in the Walawe Ganga Basin: an historical overview. In: *The South-east Dry Zone of Sri Lanka, Proceedings of a Symposium*, 29–30 April 1992. ARTI, Colombo, Sri Lanka, pp. 37–54.

Millington, P. (2000) River basin management; its role in major water infrastructure projects. Prepared as an input to the World Commission on Dams, Cape Town.

Molle, F. (2003) *Development Trajectories of River Basins: a Conceptual Framework.* Research Report No. 72. International Water Management Institute, Colombo.

Mollinga, P. (1998) On the waterfront. Water distribution, technology and agrarian change in a South Indian canal irrigation system. PhD thesis, Wageningen Agricultural University, Wageningen.

Mollinga, P. (2001) Water and politics: levels, rational choice and South Indian canal irrigation. *Futures* 33, 733–752.

Mosse, D. (1997) The symbolic making of a common property resource: history, ecology and locality in a tank-irrigated landscape in South-India. *Development and Change* 28, 467–504.

Mostert, E.N.W.M., Bouman, E., Savenije, H.H.G. and Thissen, W.A.H. (2000) River basin management and planning. In: *River Basin Management, Proceedings of the International Workshop*, The Hague, 27–29 October 1999. UNESCO, Paris.

North, D. (1990) *Institutions, Institutional Change and Economic Performance.* Cambridge University Press, Cambridge.

Ostrom, E. (1990) *Governing the Commons: the Evolution of Institutions for Collective Action.* Cambridge University Press, New York.

Ostrom, E. (1992) *Crafting Institutions for Self-governing Irrigation Systems.* Institute for Contemporary Studies Press, San Francisco.

Pointing, C. (1991) *A Green History of the World.* Penguin Books, Harmondsworth.

Rittel, H.W.J. and Webber, M.M. (1973) Dilemmas in a general theory of planning. *Policy Sciences* 4, 155–169.

Sakthivadivel, R. and Molden, D. (2001) Linking water accounting analysis to institutions: synthesis of five country studies. Paper presented at the Regional Workshop on Integrated Water Resources Management in a River Basin Context: Institutional Strategies for Improving the Productivity of Agricultural Water Management, Malang, Indonesia, 15–19 January 2001. International Water Management Institute (IWMI).

Shore, C. and Wright, S. (1997) Policy: a new field of anthropology. In: Shore, C. and Wright, S. (eds) *Anthropology of Policy: Critical Perspectives on Governance and Power.* Routledge, London, pp. 3–39.

Svendsen, M., Murray-Rust, H., Harmancioglu, N. and Alpaslan, N. (2001) Governing closing basins: the case of the Gediz River in Turkey. In: Abernethy, C. (ed.) *Intersectoral Management of River Basin.* International Water

Management Institute (IWMI), Colombo, Sri Lanka, pp. 183–214.

Technical Advisory Committee (2000) Integrated water resources management. TAC background papers no 4. Global Water Partnership, Stockholm.

Turral, H. (1998) *Hydro-logic?: Reform in Water Resources Management in Developed Countries with Major Agricultural Water Use – Lessons for Developing Nations.* Overseas Development Institute, London.

Turton, A.R. and Ohlsson, L. (2000) Water scarcity and social stability: towards a deeper understanding of the key concepts needed to manage water scarcity in developing countries. SOAS Water Issues Study Group,

Occasional Paper 17. SOAS, University of London, London.

Uphoff, N. (1986) *Local Institutional Development: An Analytical Sourcebook with Cases.* Kumarian Press, West Hartford, Connecticut.

Wester, P. and Warner, J. (2002) River basin management reconsidered. In: Turton, A. and Henwood, R. (eds) *Hydropolitics in the Developing World. A Southern African Perspective.* AWIRU, Pretoria, pp. 61–71.

Wilson, H.M. (1891) Irrigation in India. In: *12th Annual Report of the United States Geological Survey, Part 2.* USGS, Washington, DC, pp. 363–562.

2 Phases of River Basin Development: the Need for Adaptive Institutions

David Molden, R. Sakthivadivel, M. Samad and Martin Burton

2.1 Introduction

The last 50 years have witnessed enormous changes in the way people use water for productive and economic purposes. Between 1950 and 2000, the world's population increased from 2.5 to 6.0 billion (10^9). Concurrently, there was a tenfold increase in the number of large dams, from 4270 dams to over 47,000 (ICOLD, 1998). Water withdrawals have jumped from 1360 to 4000 km^3 (Shiklomanov, 2000). Agricultural withdrawals have increased from 1080 to 2600 km^3 (Shiklomanov, 2000), whereas cultivated land expansion has increased at a much slower rate from 1063 million ha to 1360 million ha. This clearly demonstrates the important role additional water has played in agricultural production. Water resources development has not only led to significant changes in land and water systems, but also in the institutions required to manage these resources.

Our working hypothesis is that changing patterns of water use require adaptive institutions for sustainable, equitable and productive management of the water resource basin wide. Institutional arrangements that are able to manage the present situation adequately are not likely to meet future needs unless they can adapt to change rapidly. This chapter presents hydraulic and institutional arguments to develop the concept of phases of river development. It uses this concept to show why adaptive institutions are necessary, what issues may arise during various phases and what the key areas of institutional focus are as river basins develop.

2.2 Hydraulic and Hydrologic Responses to Water Resources Development

To understand how institutions can more effectively manage water, it is essential to understand how the physical water resource base responds to interventions that serve human needs. River basins provide a logical unit of analysis because water flows starting from precipitation can be traced to understand the impact of these interventions. A water accounting framework is used to understand changes of water use within a basin context (Boxes 2.1 and 2.2).

With the development of hydraulic infrastructure, whether a large dam, water-harvesting structure or pump, the aim is to redirect water from its natural course to household, agricultural, industrial or other cultural use. Similarly, changes in landscapes such as from forests to farmland alter the natural hydrologic condition and redirect water flow paths. Some uses deplete water – they render it unavailable for downstream use – either by conversion of water to evaporation, or by the direction of flows into a sink such as a sea or saline aquifer. Flowing

---

Box 2.1.   Definitions.

- *Water balance and water accounting:* Water accounting relies on water balance studies for a domain bounded in space and by time. For example, we may include a portion of a basin over a year's time, bounded spatially so that runoff is captured by the sub-basin, and vertically to include the bottom of the aquifer up to the top of the vegetation canopy. A water balance analysis quantifies water flows across the boundaries including rain, evaporation, surface and subsurface inflows and outflow. Changes in storage internal to the water balance domain, such as changes in reservoir levels or groundwater levels must be considered. Essentially, water accounting divides hydrological variables of discharge, rain and evaporation into water accounting categories of process (i.e. crop evapotranspiration, domestic and industrial depletion), and non-process (evaporation and and evapotranspiration from non-agricultural land cover, committed and uncommitted outflow, and utilizable and non-utilizable outflow).
- *Diversions, depletions and recycling:* Water is diverted to various uses. Water is depleted when it is rendered unavailable for further downstream use – either through evaporation or by directing the water to sinks. Since not all water diverted to a use is depleted, some remains within the basin and is available for further use. Water recycling or reuse is prevalent in water resource systems. City effluents discharged back into river systems are often used again downstream. It is common to underestimate how much reuse exists in river systems, especially in those that are highly stressed.
- *Accounting for precipitation:* In many analyses of water resources, only the 'developed' or 'blue' water supply is considered – supplies tapped from rivers by diversion structures. In IWMI's water accounting framework, rain is considered a supply.
- *Water commitments:* All uses of water in a basin could be captured if the boundaries of a basin were defined to extend to an ideal salt–freshwater interface. Most often it is practical and useful to consider only part of a basin, but when we do this we have to make sure and define *commitments of water* to downstream uses to meet ecological or other human requirements downstream.
- *Open and closed basins:* When all available water has been allocated to various uses, we consider the basin closed. When there is water remaining in the basin to develop and allocate, we say the basin is open. In many basins, there is ample water during part of a year, and at other parts it is dry. We consider these basins to be seasonally closed.

---

water can serve several purposes – hydropower, in-stream fisheries and agriculture – and create considerable value before it is finally depleted.

Relative to other uses, food production depletes huge quantities of water. To feed one person requires that between 2000 and 5000 l of liquid water each day be converted to vapour through the biophysical process of evapotranspiration. In contrast, drinking water needs range from 2 to 5 l, and household requirements range from 20 to 500 l per person per day. As a result, food production for a growing population has significantly altered the hydrologic characteristics of many basins. Irrigation depletes water through either evapotranspiration or directing drainage flows to the sea. Rainfed agriculture depletes water that would have supported other ecosystems. Thus, there is a clear need to focus on water use in agriculture as it is a major factor in water resource and land management.

2.3 Phases of River Basin Development

Changes in water use can be illustrated by conceptualizing phases of river basin development (see Keller *et al.*, 1998; Ohlsson and Turton, 2000; Allan 2002; Molle, 2003, for similar discussions). In its most basic form, water scarcity is a situation where people have difficulties accessing water for drinking or growing crops. As a reaction to water scarcity, people tap basin water resources. Hydraulic structures, ranging from simple stone and wood diversion structures to complex dam and canal systems supply water from streamflow for drinking and industrial supplies, and for agriculture. Rainfed agriculture converts land use from its previous cover (forest, grassland) to cropland, impacting the previous hydrologic regime and ecosystems dependent on that regime.

At any particular time, the available water supply is limited by the installed

---

Box 2.2.   Water accounting categories.

- *Gross inflow* is the total amount of water flowing into the water balance domain from precipitation, surface, and subsurface sources.
- *Net inflow* is the gross inflow plus any changes in storage.
- *Water depletion* is a use or removal of water from a water basin that renders it unavailable for further use. Water depletion is a key concept for water accounting because of the interest in increasing the derived benefits per unit of water depleted. It is extremely important to distinguish water depletion from water diverted to a service or use as not all water diverted to a use is depleted. Water is depleted by four generic processes:

  - *Evaporation:* water is vaporized from surfaces or transpired by plants.
  - *Flows to sinks:* water flows into a sea, saline groundwater, or other location where it is not readily or economically recovered for reuse.
  - *Pollution:* water quality gets degraded to an extent that it is unfit for certain uses.
  - *Incorporation into a product:* through an industrial, or agricultural process such as bottling water, or incorporation of irrigation water into plant tissues.

- *Process consumption* is that amount of water diverted and depleted to produce a human intended product.
- Non-process depletion occurs when water is depleted, but not by the process for which it was intended. Non-process depletion can be either *beneficial* or *non-beneficial*.
- *Committed water* is that part of outflow from the water balance domain that is required by other uses such as downstream environmental requirements or downstream water rights.
- *Uncommitted outflow* is water that is not depleted or committed and is therefore available for a use within the domain, but flows out of the basin due to lack of storage or sufficient operational measures. Uncommitted outflow can be classified as *utilizable* or *non-utilizable*. Outflow is utilizable if by improved management of existing facilities it could be consumptively used. Non-utilizable uncommitted outflow exists when the facilities are not sufficient to capture the otherwise utilizable outflow.
- *Available water* is the net inflow minus both the amount of water set aside for committed uses and the non-utilizable uncommitted outflow. It represents the amount of water available for use at the basin, service or use levels. Available water includes process and non-process depletion, plus utilizable outflows.
- A *closed basin* is one where all available water is depleted. An *open basin* is one where there is still some uncommitted utilizable outflow.
- In a *fully committed basin*, there are no uncommitted outflows. All inflowing water is committed to various uses.

---

hydraulic structures and land placed under cultivation. When demand exceeds this available supply, one response is to provide more supply either by expanding hydraulic infrastructure or by expanding rainfed agriculture. This supply approach is ultimately limited by the amount of land and water resources within a basin, the technical and economic limits we have in abstracting this supply (it would be difficult to divert the entire Amazon), ecological thresholds beyond which ecosystems cannot sustain land and water use practices, and societal demands. The latter changes over time due to the state of a country's social development and economy.

Figure 2.1 represents a typical progression of river basin development over time with the original runoff and rainwater sources shown on the y-axis, and time on the x-axis. Over time, more water is made available for human uses from stream flow or groundwater by building structures (dams, diversions, groundwater pumps) yielding a stair-step pattern. Larger dam or diversion structures would yield a sudden jump in the amount of water made available. Smaller structures like pumps or water-harvesting devices also add more available water but in less dramatic steps. After a new dam is built, it takes time to deplete all available water by converting it into evaporation. Populations

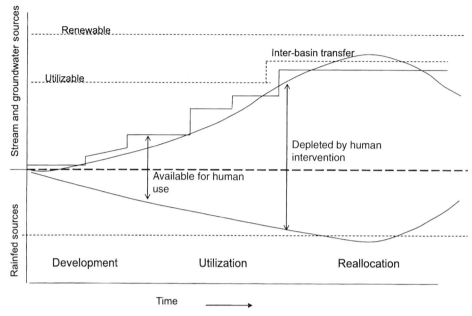

Fig. 2.1. Phases of river basin development.

grow, wealth rises and demand increases, until depletion reaches the available supply, when possibly another structure is built. Similarly, conversion of land to rainfed agriculture yields more water (directly from rain) for agriculture. Rainfed expansion continues until a limit is reached. Figure 2.1 graphically illustrates three important phases of river basin development implicit in the above discussion. The progression continues until the threshold is reached. The three phases are:

1. *Development.* In this phase, the amount of naturally occurring water is not a constraint. Rather, expansion in demand drives the construction of new infrastructure and expansion of agricultural land. Institutions are primarily engaged in expanding facilities for human use.
2. *Utilization.* Significant construction has taken place, and now the goal is to make the most out of these facilities. Water savings and improved management of water deliveries are important objectives. A common phenomenon during this phase is the growth in the amount of reuse through drains,

downstream diversions or groundwater pumps, which ultimately deplete more of the available supplies. Institutions tend to be concerned with sectoral issues such as managing irrigation water or managing drinking water supplies. The creation of the International Irrigation Management Institute, for instance, was in response to the perceived need to improve management of irrigation systems, especially new ones that were constructed in the 1950s to the 1980s.
3. *Reallocation.* When depletion approaches the potential utilizable water, there is limited scope for further development. We refer to this as a 'closed basin' (Seckler, 1996). Efforts are placed on increasing the productivity or value of every drop of water. An important means of accomplishing this is to reallocate water from lower to 'higher-value' uses. Reallocation of water to achieve both sustainability and equity among competing demands becomes a major issue. Institutions are primarily involved in reallocation, conflict resolution and regulation. When resource depletion outstrips sustainable limits, environmental restoration becomes an important agenda item.

Figure 2.1 illustrates changing depletion of water for human needs over time. Precipitation, the ultimate source, either contributes to runoff or groundwater, or is available for transpiration or evaporation providing water for life support in landscapes (rainfed sources). Over time, people tend to deplete more water by making more water available through hydraulic structures (illustrated by the stair-step patterns) and using water for agriculture, municipal or industrial uses as illustrated by the upper curve. On landscapes, evaporative water use changes by changing landscapes such as expanding rainfed agriculture as illustrated by the lower curve. The dashed lines represent thresholds, utilizable amounts that cannot be exceeded over long periods. As human depletion increases, river basins pass through phases of development, utilization and reallocation (Box 2.1).

In many closed or closing basins, supply side responses prevail, and water depletion exceeds the potentially available resource – in the long run, a non-sustainable situation. For example, in the North China Plains (CCAP and CAAS, 2000; Kendy *et al.*, 2003) and in northwest India (Molden *et al.*, 2001), intensive groundwater use is lowering aquifers to unsustainable levels. As another example, people will 'mine water' from important environmental reserves by destroying wetlands, or depleting water before it reaches its destination such as Lake Chapala (Wester, this volume) or the Aral Sea.

In the case of closed or closing basins, inter-basin transfers can provide relief. A south to north diversion of water from the Yangtze to the Yellow river basin could provide some relief to the highly stressed North China Plains. The final step of the stair-step pattern on the *available resource* and *utilizable resource* lines in Fig. 2.1 represents an inter-basin transfer. There are other responses though – such as limiting or managing demand for water, or reducing agricultural use of water. National policy responses include providing other non-agricultural employment or importing 'virtual' water (Allan, 1998) in the form of trade in commodities into the basin. How societies

respond to the crisis of overexploited resources is one of the critical water resource management questions of our times.

In overly stressed basins, restoration and sustainable use of river systems becomes a key agenda item. In the western USA, for example, dams are being dismantled and an increasing amount of the available supply is allocated to environmental uses. In California, 46.5% of supplies were allocated to environmental uses compared to 11% for urban uses and 42.5% for agricultural uses in 1995 (Svendsen, this volume). In Australia, the Government of New South Wales recently reduced allocations to irrigation by 10% so that allocations to the environment could be increased (Hatton MacDonald and Young, 2000). In other situations, such as Pakistan or the North China Plains, concerns over water-based food security and rural employment would make a reallocation of water out of agriculture a very difficult political choice. The danger though is that the ecosystems that support agriculture may collapse through salinization, loss of soil fertility or groundwater depletion before other rural income generating and food security solutions are found.

Pathways of basin development vary from basin to basin. For example, available supplies may actually decline over time, intersectoral reallocation may come into play early or interbasin transfers may occur early in the development phase. Therefore, it should not be construed that following the pathway of Fig. 2.1 is ideal, or that a basin in the reallocation phase is better than one in the development stage. To understand basin dynamics and response options for the future, it is important to understand development pathways.

Similarly to river basins, locally managed irrigation has its own development life cycle, which has to adjust to the external environment within which it is set. Irrigation schemes that are transferred while the river basin is in an early development stage have fewer external issues to deal with. By contrast, those transferred in a latter stage of river basin development need to focus both internally on the use of water within the system, as well as externally to ensure a continued share of the available scarce water resource. The manner in which these locally managed organizations react under these different conditions is dealt with in Chapter 13. Nine essential supporting tasks are identified to support locally managed irrigation (see Chapter 1). The following is a discussion of how these essential tasks take shape and gain or lose importance according to the phase of river basin development.

2.4 Different Phases – Different Needs

Institutional concerns differ depending on the phase of basin development. These concerns may exist at all times, but their importance or emphasis may change over time as illustrated in Table 2.1. Institutions have to adapt to meet these changing concerns. For example, interest may shift from constructing facilities to provide supply to better management of supplies. As basin closure approaches, however, demand management becomes a critical concern. At any phase, though, there could be overlap. For example, even in the development phase, managing demand could be a concern especially when depletion approaches available supplies. The following sections

describe how essential institutional tasks vary as per phases of development.

2.5 Essential Institutional Tasks

2.5.1 Planning, construction and maintenance

In the development phase, planning for and implementing infrastructure construction plays a dominant role. In modern times, organizations such as the United States Bureau of Reclamation, the Mahaweli Authority and the DSI in Turkey were set up to build major dams, canals and, sometimes, drinking water treatment and wastewater plants. Initially, these agencies were dominated by civil engineers who had the important job of getting high-quality work done quickly. These tasks are most easily carried out by single organizations instead of several organizations because of problems of coordination and high transaction costs if many are involved.

During the utilization phase, a different set of construction and maintenance activities gain prominence. Maintenance becomes a concern to keep facilities in operating order, and rehabilitation is warranted to bring systems back to operating order. Measurement and regulation structures are often

Table 2.1. Various dominant characteristics and concerns at different phases of river basin development.

Characteristic	Development	Utilization	Reallocation
Dominant activity	Construction	Managing supplies	Managing demand
Fraction of utilizable flow depleted	Low (0–0.4)	Medium (0.4–0.7)	High (0.4–>1.0)
Value of water	Low value of water	Increasing value of water	High value of water
Infrastructure	Installing new structures	Modernization/ rehabilitation	Measurement, regulating
Groundwater	Utilizing groundwater	Conjunctive management	Regulating groundwater
Pollution	Diluting pollution	Emerging pollution/salinity	Cleaning up pollution
Conflicts	Fewer water conflicts	Within system conflicts	Cross sectoral conflicts
Water scarcity	Economic water scarcity	Institutional water scarcity	Physical water scarcity
Data	Water data – perceived importance less	System water delivery data important	Basin water accounting data important
Water-poverty concerns	Including/excluding poor in development of facilities	Including poor in O and M decision making	Loss of access to water by poor

added to assist the operation of facilities. Modernization is often pursued to meet the changing expectations of users or societies.

During the reallocation phase, restoration of environmental damage, clean-up of pollution, or placing procedures to limit or reduce demand tend to capture the attention of planners. Large and complex infrastructure development projects, such as inter-basin water transfers, are often sought to solve water problems. Decentralized construction efforts such as the installation of pumps, small reservoirs or water-harvesting structures play a significant role in the development and management of water resources throughout all the phases of development. Often these are not centrally planned but individual or community responses to water scarcity, with private or community investments. One small structure may not impact much on other uses, but the cumulative sum of thousands of such small works can quickly push a basin up the development curve. In China and India, for example, the amount of area served by groundwater irrigation has surpassed the land irrigated by large-scale surface schemes. Therefore, an institutional challenge is to incorporate these efforts into broader basin management schemes while recognizing the value of these locally driven initiatives.

2.5.2 Water allocation and distribution

Important allocation decisions are made right at the development phase. The decision of where to place a reservoir or who gets subsidized support for pumps is basically a water allocation decision. Some people will benefit, whereas others will be excluded from access to water.

The utilization phase is characterized by operationalizing *within*-system allocation and distribution procedures. Often these procedures are ill-conceived during planning and construction, or the institutions meant to carry out these tasks are ill-equipped to do so. Allocation and operating distribution systems takes a different set of skills than building large infrastructure, yet it is often the same people initially in charge of construction who make these operational decisions.

In the reallocation phase, cross-sector and between-system allocation and reallocation become increasingly dominant concerns. For example, now a major concern of Turkey's DSI is formally to allocate and distribute bulk supplies of water across uses and sectors.

2.5.3 Monitoring and enforcing water quality and protecting the environment

During the development phase, the development of water infrastructure and rain-fed land significantly change landscapes, affecting the environment. The changes are often large and detrimental and include changing the hydrologic regime of rivers, impounding large bodies of water and drying up wetlands. During the utilization phase, water use and depletion intensify further, removing water that may have environmental functions. Throughout the early phases of development, dilution can be sufficient to solve pollution problems. During reallocation phases, dilution is not an option, because there simply is not enough water. Clean-up at the source becomes increasingly critical. In the development phase, institutions initially must be concerned with the impact of large infrastructure development such as relocating people, or removing large areas of habitat. Later, in utilization and reallocation phases, problems of pollution, salinization or over-withdrawal of resources become more important.

2.5.4 Protecting against water disasters

Floods and droughts are key trigger mechanisms in changing the development pathways of river basins. While long-term planning and overall economic development of countries play an important role in determining the path of river basin development, extreme events have a very important role in shaping water development trajectories. For example, a devastating flood or drought can provide enough political backing for the construction of a new reservoir. Protecting against these disasters is important at all phases of development, but takes on different forms. Initially, people construct storage facilities. Later operation becomes more critical, and more sophisticated early-warning systems and operational systems developed. Sometimes, droughts spur the development of more groundwater resources, pushing water resource use beyond sustainable limits.

2.6 The Outcomes: Water Scarcity, Value of Water and Poverty

2.6.1 Water scarcity

Scarcity takes on different characteristics during various phases of development. People feel water scarcity if they cannot obtain water, even though it may be plentiful in nature. Construction of facilities provides access to water, and relieves water scarcity. In the development phase, *economic water scarcity* (IWMI 2000) is common. This is a condition where financial or human resources limit access to water.

However, even with facilities to provide access, scarcity can exist. A condition of *institutional water scarcity*[1] exists when laws, customs, traditions and other social barriers, or organizations deny certain groups of people easy access to water or prevent the equal distribution water to all, leaving some people water scarce. A persisting head–tail problem where overall water is adequate in an irrigation system is an example of this type of scarcity. Another example is when customs or laws restrict access by a certain group of people who are deemed to be in the lower strata of the social hierarchy.

During the reallocation phase, the absolute supply of the physical resource is limited. This is a condition of *physical water*

scarcity. In many closed or closing basins, such as those in the North China Plains, the Gediz basin or the Colorado River, physical scarcity limits the supply of additional water. It is important to note that even with physical scarcity, it is possible to manage institutions so that individuals do not suffer from lack of access of adequate or good-quality water, though changes in water use, or even livelihoods, may be required.

2.6.2 The value of water

The value of water tends to increase through the different phases of development. Early, when water is plentiful, water has a relatively low value, but when a basin closes and demands for a scarce resource intensify, the value of water can rise dramatically. Accordingly, in the early phases of development, concerns are centred on developing a supply of low-valued water, while later in the development process, managing demand prevails. Agriculture and environmental uses are commonly perceived as low-valued uses, while industry, hydropower and domestic uses have a higher economic value. A common phenomenon is the initial reallocation of water from environmental uses to agriculture, then from agriculture to cities. Often wealthier countries place increasing value of water in environmental uses, and reallocate water back to these uses in the reallocation phase. When low-valued water is plentiful, conflicts can be mitigated with more supplies. As supplies become limiting, the potential for conflict increases.

2.6.3 Water and poverty

In each phase of development, water scarcity has important implications for poverty (van Koppen, 2000). During the development phase, a key water and poverty concern is to target the poor in water

---

[1]   First described by M. Samad.

development efforts. An important consideration is to identify the beneficiaries. Will infrastructure benefit poor people? Will more powerful people capture the benefits? Providing water to a limited number of people in a society with pronounced social differentiation will aggravate inequities. Poverty can be exacerbated or relieved depending on the planning decision made.

During the utilization phase, having a voice in allocation and receiving a fair share is a problem of the poor. Even though conveyance structures exist, the way water is managed may not meet the needs of the poor. If the rich or upstream users are able to capture water more readily than tail-enders or the poor, the poor suffer. For example, in Kathmandu, homeowners often pump water from city pipelines sucking out cheap supplies, and forcing the less fortunate to purchase high-priced tanker water.

During the reallocation phase, water is reallocated amongst sectors and people. Maintaining access to water is a key problem for the poor during this phase. People with financial or political power tend to receive more benefits from water. If water moves away from agriculture to higher-valued cities and industries, will the poor and less powerful be able to maintain their right or access to water, or will they find employment in other sectors? Will poor people be able to capture the economic gains when water moves to higher-valued uses?

2.7 Adaptive Institutions

If, early in the development phase, attempts are made to design an institutional framework that deals with all these issues – pollution, poverty, allocation, regulation, construction – it is not likely that success would be achieved across the board. At certain phases of development, some of these concerns take precedence over others. Well-functioning construction agencies have an important role in safe and sound construction, but using an institutional set-up designed to manage construction and also to manage water delivery is not likely to be effective. Different sets of rules and skills are required. Similarly, those who manage service delivery are probably not equipped to regulate allocation and pollution of resources when these problems emerge.

The implication is that water resource management institutions must adapt to meet different challenges as patterns of water use change. Common water problems are seen because agencies do not change fast enough to adapt to changing needs. Institutional set-ups then must have the means to recognize when important changes take place, and the means to adapt when necessary.

2.8 Development Pathways

At early phases of development, future choices about water are many more than when a basin is at the reallocation phase. Near basin closure, many difficult-to-reverse decisions about allocation and infrastructure development have already been made. The agenda is to solve complex problems of reallocation and degradation. Ecosystem restoration is an important agenda in some locations with controlled water being reallocated back to natural uses.

In other basins, with high rural populations and few off-farm employment opportunities, maintaining intensive use of water in agriculture seems the only option until further alternative economic development takes place. A solution for the problem of groundwater depletion in the North China Plains would be simply to curtail the amount of agriculture. At the moment, this is almost unthinkable as possibly millions of people dependent on agriculture would lose an important means of livelihood and food security. However, unless solutions in other economic sectors or within the water and agriculture sector can be found, the use of water for agriculture may cease to be an option because of its unsustainable use. The impact on livelihoods would be the same.

Fortunately, at early phases, several key choices remain: how much more rainfed and irrigated agriculture; how should this be done through dams, groundwater, water harvesting; what are desirable landscape patterns and what are associated water consequences; how can additional water supplies be allocated to the rural poor; and how much water to commit to ecological uses. Societies may choose a path of intensive agricultural development in the hope of economic development and later restoration. The risk of this approach is ecological collapse, or very high costs due to poor health, and clean-up of degraded environment. Lundqvist (1998) refers to this situation as 'hydrocide'. Another development pathway would rely on using less water, particularly in agriculture, and increasing the value or productivity from each drop used. The associated risk is not using water to enhance income and food security for local poor populations.

All too often decisions are made in response to a natural or political crisis without consideration of long-term goals. The available choices at early phases of development are not very well thought out. A question is whether institutional arrangements can evolve and have the resilience to solve short-term problems, in light of key long-term growth and sustainability goals.

2.9 Summary and Conclusions

The growing recognition of a river basin as the most appropriate unit for the development and management of water resources has prompted the search for appropriate institutional arrangements for river basin management. This chapter has argued that institutional requirements differ with the different phases of development of the river basin. Institutional designs must be set up to adapt to these changes. Thus, an understanding of basin development pathways is crucial in understanding or formulating institutional arrangements for river basin management. This chapter has outlined a framework to chart phases of river basin development based on hydraulic and hydrologic changes.

We demonstrated that as a river basin progresses from an 'open' to a 'closed' basin,

three idealized phases can be distinguished: development, utilization and reallocation. These are not mutually exclusive and some overlap of functions may occur. At the early stages of development, institutional arrangements focus on a single or very limited set of objectives, typically involving developing infrastructure to supply water. Later, more concern is placed on managing water within various sectors. When demand outstrips supplies because of more utilization, competition increases, the value of water increases, and a host of other issues including environmental concerns, pollution and groundwater overdraft may arise. Over time, there is an increasing need to deal with multiple functions that require complex institutional arrangements that involve several organizations, and that function in the realm of a broader and often conflicting set of national objectives. Thus, institutions should be conceived as dynamic entities that need to cater to different management demands as water use changes over time. A key feature of an effective institution will be its ability to adapt to changing needs.

Acknowledgements

The authors appreciate the constructive suggestions by Mats Lannerstad, Gina Castillo and Mark Svendsen.

References

Allan, T. (1998) Moving water to satisfy uneven global needs: 'trading' water as an alternative to engineering it. *ICID Journal* 47, 1–8.

Allan, T. (2002) Water resources in semi-arid regions: real deficits and economically invisible and politically silent solutions. In Turton, A. and Henwood, R. (eds) *Hydropolitics in the developing world: a Southern African perspective*. African Water Issues Research Unit, University of Pretoria, Pretoria.

CCAP and CAAS (2000) *Water Accounting for Fuyang River Basin*. 2000. Draft Interim Report prepared by the Chinese Center for Agricultural Policy and the Chinese Academy of Agricultural Sciences for the Five Country Regional Study on Development of Effective Water Management Institutions.

Hatton MacDonald, D. and Young, M. (2000) Institutional arrangements in the Murray–Darling River Basin. In Abernethy, C.L. (ed.) *Intersectoral Management of River Basins: Proceedings of an International Workshop on Integrated Water Management in Water-Stressed River Basins in Developing Countries: Strategies for Poverty Alleviation and Agricultural Growth*, Loskop Dam, South Africa, 16–21 October 2000. IWMI/DSE, Colombo, Sri Lanka.

ICOLD (International Commission on Large Dams) (1998) *World Register of Dams*. ICOLD, Paris.

IWMI (2000) *World's Water Supply and Demand 1995–2025*. IWMI, Colombo, Sri Lanka.

Keller, J., Keller, A. and Davids, G. (1998) River basin development phases and implications of closure. *Journal of Applied Irrigation Science* 33, 145–163.

Kendy, E., Molden, D.J., Steenhuis, T.S. and Liu, C. (2003) Policies drain the North China Plain: agricultural policy and ground-water depletion. In: *Luancheng County, Hebei Province, 1949–2000*. IWMI Research Report, forthcoming.

Lundqvist, J. (1998) Avert looming hydrocide. *Ambio* 27, 428–433.

Molden, D., Sakthivadivel, R. and Samad, M. (2001) Accounting for changes in water use and the need for institutional adaptation. In Abernethy, C.L. (ed.) *Intersectoral Management of River Basins: Proceedings of an international workshop on Integrated Water Management in Water-Stressed River Basins in Developing Countries: Strategies for Poverty Alleviation and Agricultural Growth*, Loskop Dam, South Africa, 16–21 October 2000. IWMI/DSE, Colombo, Sri Lanka, pp. 73–87.

Molle, F. (2003) *Development Trajectories of River Basins: a Conceptual Framework*. IWMI Research Report, forthcoming.

Ohlsson, L. and Turton, A. (2000) The turning of a screw. Social resource scarcity as a bottle-neck in adaptation to water scarcity. SAS Water Issues Study Group Occasional Paper 19. University of London, London.

Seckler, D. (1996) *The New Era of Water Resources Management: From 'Dry' to 'Wet' Water Savings*. IWMI Research Report 1, Colombo, Sri Lanka.

Shiklomanov, I.A. (2000) Appraisal and assessment of world water resources. *Water International* 25, 11–32.

van Koppen, B. (2000) *From Bucket to Basin: Managing River Basins to Alleviate Water Deprivation*. International Irrigation Management Institute, Colombo, Sri Lanka.

3 Limits to Leapfrogging: Issues in Transposing Successful River Basin Management Institutions in the Developing World

Tushaar Shah, Ian Makin and R. Sakthivadivel

3.1 Backdrop

Management becomes important as a productive resource becomes scarce, and there is hardly a situation in which this is truer than in the case of the water resource. For a long time now, water policies of many emerging nations have been focused on developing the resource, and optimizing was directed at the efficiency of water infrastructure rather than water itself. As water has become increasingly scarce, optimizing is now being increasingly directed to improving the productivity of water. Increasingly, the river basin is emerging as the unit of management of land, water and other natural resources in an integrated fashion. Many developed countries such as the USA, France and Australia have evolved highly advanced and resilient institutional regimes for integrated river basin management (IRBM); but this has taken decades – in Europe and the USA, centuries – of gradual change to evolve. An issue that has held great appeal to policy makers and social researchers is this. Is it necessary that developing countries in Asia and Africa take all that long in crafting such institutional regimes? Or might it be possible for them to do an 'institutional leap-frog', as it were, to

a stage at which developed-country basin institutions are today?

A textbook case of institutional reform for IRBM in recent times has been the Murray–Darling basin in Australia, where sweeping changes have been made and enforced since 1990, and transferring the lessons of success in IRBM – from Murray–Darling to Mahaweli and Mekong – has emerged as a growth industry.

This chapter attempts a broad-brush approach to understanding the material differences between the contexts of the developed-country river basins from where institutional models emerge and the developing-country river basins in which these are applied. The idea is not to undermine the significance of the lessons from success but to emphasize the need for sagacity and critical analysis in assessing what will work and what will not, given the differences in context. The phrase 'institutional change' is used to describe how communities, government and society change recurrent patterns of behaviour and interactions in coping with water scarcity and its socio-ecological ill-effects. It involves understanding laws and rule making, roles, policies and institutional arrangements at different levels. The overarching premise is that the effectiveness of a pattern of institutional

©CAB International 2005. *Irrigation and River Basin Management*
(ed. M. Svendsen)

development is determined by at least four realities of a river basin: hydrogeology, demography, socioeconomics and the organization of the water sector. By implication, institutional arrangements that have proved effective within one set of these realities may require major adaptation before they become appropriate to the needs of a river basin context defined by an alternative set of realities.

IRBM is a powerful idiom, and will increasingly dominate natural resource management discussion in the developed as well as developing world. In its broadest sense, a basin or catchment is visualized as

> an inter-connected machine or system which transforms natural inputs of solar energy, atmospheric precipitation, nutrients and other environmental factors, along with man-made inputs of labour, capital, materials and energy, into output products such as food, fibre, timber, building materials, fuels, minerals, natural vegetation and wildlife, recreational and aesthetic amenities, buildings and development sites, as well as water in desirable quality and quantity. (Burton, 1986, cited in Hu, 1999, p. 324)

River basin management (RBM) as a notion goes far beyond traditional land and water management and

> includes significant parts of land-use planning, agricultural policy and erosion control, environment management and other policy areas. It covers all human activities that use or affect fresh water systems. To put it briefly, RBM is the management of water systems as part of the broader natural environment and in relation to their socioeconomic environment. (Mostert *et al.*, 1999, p. 3)

Institutional discussions on IRBM have tended invariably – and probably erroneously – to gravitate around three models of strategic organizations for managing river basins:

● The *hydrological model*, in which a river basin organization/authority, cutting across administrative boundaries, takes over all charge of water resource management;

● The *administrative model*, prevailing in many developing countries, in which water management is the responsibility of territorial organizations unrelated to hydrological boundaries; and

● *Coordinating mechanisms* superimposed on the administrative organizations to achieve basin management goals.

Each has advantages and disadvantages. The hydrological model effectively deals with upstream–downstream issues that the administrative organization is generally unable to deal with. However, hydrological organizations tend typically to focus on water and overlook land management issues. River basin commissions, as a hybrid, might combine the advantages of both, but at least in the developing-country context, they often command little authority, and are therefore confined to lowest-common-denominator solutions (Mostert *et al.*, 1999). In many developing countries today, institutional reform for RBM is confined almost wholly to the creation of the basin-level organization – the implicit assumption being that mere formation of the appropriate organization will result in IRBM, an assumption whose validity has been repeatedly refuted.

In the developed world, the discussion has been much broader and has veered around initiatives in four aspects of natural resource governance:

1. Some mechanism for basin level negotiation and coordination fortified with adequate authority and resources, and a broad mandate considered appropriate to the basin's context;
2. Legal and regulatory reform;
3. Redesigning economic instruments of policy (transfer prices, taxes, subsidies) in harmony with national policy goals;
4. Redesign of economic institutions (including utilities, service providers, property rights; water markets, irrigation management transfer to user organizations).

Countries like the USA have achieved, over long periods, high levels of integration even without a central basin organization (e.g. Svendsen, Chapter 7, this volume).

3.2 Applying the Lessons of the Murray–Darling to the Developing World

The Murray–Darling River system, as a recent case of accelerated institutional reform, has emerged as a model of institutional structure for IRBM. The basin encompasses over 75% of the state of New South Wales, 56% of the state of Victoria, all of the Australian Capital Territory, and small parts of Queensland and South Australia, a vast region of the southeastern part of the continent. Already, several case studies of the Murray–Darling are available, and it is not our intention to review these. In brief, the institutional innovations of the Murray–Darling basin management regime include the following:

1. The Murray–Darling Ministerial Council as the top-level policy-making and coordinating mechanism, the Murray–Darling Basin Commission as the operating organization, and several Catchment Management Agencies that are responsible for day-to-day management of water;
2. A system of permits for diversions that encompasses all uses except the water needed for domestic use, livestock production and irrigation of up to 2 ha which are recognized as prior rights (MacDonald and Young, 2000, p. 10), and exempted from legal as well as permit systems;
3. An effective cap on water diversions at 1993–1994 levels of development to ensure adequate environmental supplies, accompanied by a system of volumetric licensing to users that raises the scope for large-scale water trade across states and sectors;
4. Consumption based, full-cost-recovery pricing (MacDonald and Young, 2000, p. 14);
5. A system of 'salinity credits' that permits trade in salinity;
6. Explicit mechanisms for water allocation for environmental needs;
7. A legal regime that separates water rights from land rights;

8. Privatization of service providers such as Murray Irrigation Ltd and Victoria's Rural Water Corporation (Malano et al., 1999).

The Murray–Darling RBM regime clearly represents a highly evolved form of institutional arrangement and effectively addresses all major problems that a mature river basin would face. As alluded to earlier, exploring whether developed-country basin institutions – particularly, Murray–Darling experience – can be replicated in a developing-country context has fascinated many researchers in recent years. An entire issue of *Water International* (vol. 24, no. 4, 1999) was devoted to it in 1999. The results of these investigations have not been very encouraging. For example, Hu explored the applicability of Murray–Darling experience in the Chinese context and concluded negatively because of: (i) difficulty of coordinating authorities at different levels; (ii) unclear ownership of resources; (iii) small farming scales; and (iv) poor education of resource users (Hu, 1999, p. 323).

In a similar vein, Malano et al. (1999) ask: 'Can Australian experiences be transferred to Vietnam?' Their conclusion is less emphatic than Hu's, but all their evidence suggests that it will be long before Vietnam becomes really ready for the Murray–Darling prescription, and that 'context, hydrological and socioeconomic, defines the detail and balance that is required . . .' (p. 313). The new water law of Vietnam contains provisions to adopt an integrated river basin approach. The Ministry of Agriculture and Rural Development, which is at present in charge of water, does not relish the responsibility of IRBM. The progress in stakeholder participation, another Murray–Darling prescription, has been slow to say the least. Farmers view irrigation provision as a government responsibility; even so, irrigation charges in Vietnam are high by Asian standards. Yet, presumably under donor pressure, the government tried to eliminate irrigation subsidies, but this was followed by massive popular unrest in 1998, whereupon, the Government had to restore the subsidies.

Can the Australian success in enforcing the 'user pays' principle be transferred to the

Solomon Islands? Hunt explored this issue in a recent study and concluded that such transfer 'is not sustainably viable' on account of huge differences in political structures, national priorities, living standards, cultural traits, technological development, literacy levels, financial and infrastructure growth, and change-management competency. All these differences result in the absence of what Hunt calls a 'contextual fit' between the policy development and the respective policy application environment (Hunt, 1999, p. 302).

'If there is any conclusion that springs from a comparative study of river systems, it is that no two are the same' (Gilbert White, cited in Jacobs, 1999). Each river basin must differ from any other in a thousand respects, but that does not mean that lessons of success in one are of no value to another. It does mean though that uncritical 'copycat' replication of successful institutional models – either by enthusiastic national governments or at the behest of enthusiastic donors – is a sure formula for failure. The history of institutional reform in developing-country water sectors is dotted with failures of such copycat reform.

IRBM is not a new idea, even in developing countries. India tried to transpose the TVA (Tennessee Valley Authority) model tried in the USA by constituting the Damodar Valley Authority, which was a resounding failure. Catchment management committees were established in China way back in the 1950s in some of the major river basins such as the Yangtse and Yellow River, to plan and exploit water resources, generate electricity, mitigate flood damage and provide facilities for navigation (Hu, 1999, p. 327). However, all these institutions shed their broad agenda and ended up focusing on irrigation, the purpose that was most central to their domains at those times.

In Sri Lanka, a Water Resources Board was established as early as 1964 to promote integrated water resources planning, river basin and trans-basin development and to tackle water pollution. However, the Board never worked on its broad mandate and, instead, took to hydrological investigations and drilling tubewells.[1] Such examples can be multiplied easily. The point is, in taking useful lessons from success cases for making meaningful reform in developing countries, it is important to understand critical differences between the two worlds that affect what will work and what will not. We propose that, in assessing the applicability of institutional innovations, it is critical to take into account four types of *material* differences between the developed- and developing-country realities:

1. Hydrology and climate;
2. Demographics;
3. Organization of the water sector;
4. Socioeconomics.

We briefly outline these material differences in the following sections.

3.3 Hydrology

Historically, agriculture advanced early in arid climates such as those of Egypt and Iraq, but industrial development began early in the temperate and humid climates of Europe, North America and Japan. Arid areas where significant wealth creation and accumulation has occurred, such as in West Asia, are typically rich in mineral and oil resources. As of today, however, the bulk of the developing world, where rainfall tends to be low and water scarcity is a major emerging constraint to progress, is in the arid or semiarid[2] parts of the world. Figures 3.1 and 3.2, showing the global distributions of mean annual rainfall and

---

[1]   Another round of reform has just begun in Sri Lanka. In 1990, a draft law made provision for bulk water allocation and the establishment of a National Water Resources Council to do what the Water Resources Board could not. However, the law could be submitted to the Parliament only in 1995 for the lack of consensus within the cabinet as well as amongst the myriad agencies dealing with water (Birch and Taylor, 1999, p. 331).
[2]   Referring to regions like India and West Africa, which are humid for a small part of the year but arid during the rest of the year.

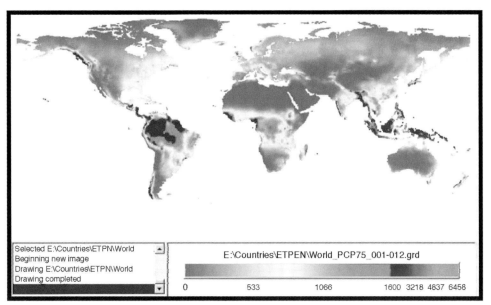

Fig. 3.1. Global distribution of mean annual rainfall.

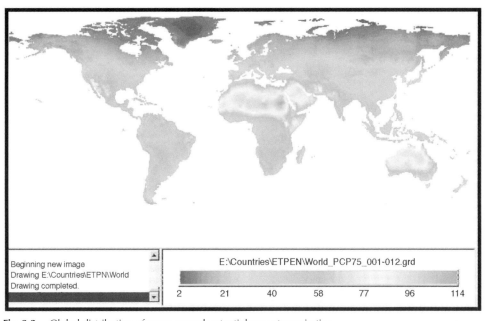

Fig. 3.2. Global distribution of mean annual potential evapotranspiration.

potential evapotranspiration, help to illustrate some major climatic differences between developed countries (mostly in the temperate latitudes) and developing countries (mostly in the tropical and subtropical regions).

Sutcliffe (1995) pointed out that developing countries also happen to be concentrated in parts of the world with more extreme climates when compared to the regions occupied by today's developed countries. Figure 3.1 illustrates the variation in annual rainfall. India, for example, receives almost all of its annual rainfall in less than 100 h of torrential downpours during June–October, and its challenge is to save enough of it from evapotranspiration to last from October until April–May, the months that mark the period of highest water stress. Botswana receives all of its 350–500 mm of rainfall during November–March, the period which also coincides with the highest evaporation, resulting in little or no runoff (Sutcliffe, 1995, p. 69).

Humid areas typically have higher stream densities than are found in the arid and semiarid areas, which means that, *ceteris paribus*, a higher proportion of precipitation in the arid and semiarid areas runs off in sheet flow before forming into streams, and is thereby subject to higher evapotranspiration losses (Fig. 3.2). Other things also are not quite the same. The tropical developing world – especially, South Asia and much of Africa – has higher mean temperatures for more of the year than the developed world, and, for equivalent levels of precipitation, runoff and the need for irrigation tend to be greater in arid and semiarid areas than in humid areas (Sutcliffe, 1995, p. 64).

Climate and hydrological conditions, combined with demography (discussed in the following section), explain why decentralized institutions for water management have historically evolved in many parts of the developing world. The profusion of small tanks in India's southern peninsula and Sri Lanka can be viewed as the response of communities in the catchment areas to stake their claim on rainfall. Even today, one collective maintenance task carried out by many South Indian tank communities before the start of the monsoon is cleaning and deepening the channels that feed rainwater runoff to their tanks. Village people here recognize that if they do not capture runoff in artificial streams, most of it will be lost before it reaches their tanks.

3.4 Demographics

Many parts of the developed world have extreme climates too; however, over time, population and urbanization in these parts have tended to concentrate in wet areas or on downstream reaches of rivers near coastal areas, where water can be supplied through large-scale diversion structures. As Fig. 3.3 shows, except in Europe, most of

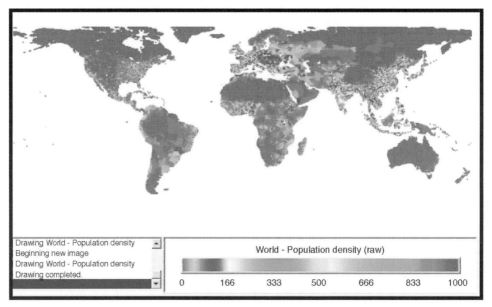

Fig. 3.3. Global distribution of population density.

the developed countries have low population densities throughout, with urban agglomerations near the coasts and rural population along rivers or irrigation systems. Here, the competition is for large accumulated bodies of 'diverted water'. Since catchment areas have relatively sparse populations, the downstream water-harvesting structures have large catchment areas that are virtually free from competition.

However, this is not the case in some of the most densely populated regions of the world. In India, for instance, population density is high – approaching 600 persons per km^2 in the water-rich Ganga basin, and seldom less than 350–400 even in semiarid Western India and hard-rock peninsular India. Population density is high both upstream and downstream of dams. The same is true for much of China. The North China plains have much less water than South China, but their population density is around the same. One might argue that the cause of intensive groundwater development in South Asia and China is that most people here cannot be downstream of large dams. By sinking tubewells, people upstream are, in a sense, challenging the basic inequity inherent in the pattern of large irrigation projects that usurp the rainfall precipitation of populous upstream catchment areas to bequeath it to a small number of canal irrigators.

All these factors have had implications for the kind of water institutions that have evolved historically in the developed and developing world. For example, the system of rights based on riparian doctrine and on the doctrine of prior appropriation is alien to the cultures of many developing countries because the largest majority, by far, depend upon rainfall and local water-harvesting and storage structures. Riparian rights or prior allocation become operative only along the streams and rivers, where the bulk of the irrigators and water users tend to be concentrated in countries like the USA or Australia, but these make no sense, for example, for some 20 million persons pumping groundwater in South Asia, or the communities that use over 300,000 tanks in South India or 7 million ponds in China.

Because large proportions of the population in the developing world depend upon rain and on local storage, the people's notions of ownership and rights relate more easily to precipitation than large-scale public diversions. Egypt gets less than 10% of its water from rainfall, yet Egyptians consider the rainwater to be truly their own. In Asia, where population densities are commonly as high in the catchment areas of the basin as along the stream and river channels, the implicit primacy of the right of communities over precipitation rather than over diversions is widely accepted. Indeed, in recent years, a popular slogan in Western India is 'rain on your roof, stays in your house; rain on your field stays in your field; and rain in your village stays in your village'. In the Western countries, upstream–downstream conflicts are important because most water users think of users upstream as their rivals. In the World Water Forum that met at The Hague in March 2000, the slogan that the Delhi-based Centre for Science and Environment popularized was 'everyone lives downstream', which is eminently sensible if all or a majority of people in a basin depend for their water needs directly upon rainfall. The IRBM discussion talks very little of the enormous amount of work on farming in the semiarid tropics done by national and international centres such as ICRISAT.[3] As the Global Water Partnership (2000, p. 25) notes, 'Most water management, including the literature on IWRM, tends to focus on the 'blue water',[4] thus neglecting rain and soil-water management. Management of 'green water' flows holds significant potential for water savings.' This is because there is little real 'dryland farming' of the Indian and West African variety in the developed

---

[3] International Crop Research Institute for the Semi-Arid Tropics.
[4] 'Blue water' is water existing in bodies such as rivers or lakes, or pumped from aquifers. 'Green water' is soil water extracted and transpired by plants.

world; but making the best use of soil moisture is a critical issue in many African and Asian countries. Europe, Canada, New Zealand and the USA do have rainfed farming; but this is not quite the same as dry farming in western Rajasthan or sub-Saharan Africa; in many of these countries, favourable rainfall and climate conditions result in favourable soil-moisture regimes that make irrigation unnecessary for growing good crops.

The conventional notion of irrigation is one of controlled supply of water to meet the full scientific requirements of plants precisely when needed, but the irrigation that is most widely practised in South Asia and amongst smallholder farmers in Africa is supplemental irrigation designed to increase the productivity of 'green water'. Green water is the precipitation used directly for crop production and thus 'lost' in evaporation; blue water pumped out from wells is as important in South Asia and North China as the part that flows into rivers and canal systems.[5] This is quite unlike the situation in many developed-country river basins. In these, the bulk of economic water demands have been met from development of blue surface water and where, with the closure of these basins, the focus of basin management is on raising the productivity of blue surface water, largely without regard to green water.

Uniformly high population density, combined with unhelpful climate and hydrology, has a profound impact on the objectives of water management in developing-country river basins. In recent years, IWMI's water accounting work (Molden and Sakthivadivel, 1999, pp. 58–60) has made much contribution to understanding water productivity in the basin context.[6] Although IWMI's focus has been on productivity of water in agriculture, the framework can be easily generalized to develop a notion of basin-level water productivity in terms of a social welfare function for all stakeholders in a river basin constituting a basin community. Under this broad concept:

> Basin welfare productivity of water = Basin welfare/Available water.

Water productivity understood thus could be enhanced by:

1. Enhancing productivity in each use; and
2. Constantly reallocating water amongst alternative uses – irrigation, domestic, industrial and environmental – so that the marginal contribution to overall welfare by water allocated to all uses remains equal.

Using the IWMI water accounting framework, this welfare productivity measure can be written in several alternative ways to highlight the importance of different water use strategies,[7] but for highlighting the difference between developed and developing world, a useful way to write the welfare productivity ratio is:

> Basin welfare/Available water = [Basin welfare/Diversions] × [Diversions/Available water].

In relatively water-abundant humid regions, with low population density in the catchment areas and dense human settlements near the coasts and along rivers, RBM seeks to maximize basin welfare productivity by increasing the term [Basin welfare/Diversions]. Allocation of diverted water amongst alternative uses is a crucial function in basin-level water management in such conditions. Here, reservoirs have large free catchments, and evapotranspiration in catchment areas is often not high; therefore, the need for active human intervention to maximize [Diversions/Available water] is not great.

---

[5]   This distinction between green and blue water is extremely important for developing countries in the semiarid tropics. Terrestrial ecosystems are green-water dependent; aquatic ecosystems are blue-water dependent (GWP, 2000, p. 24).
[6]   Standard definitions used in IWMI water accounting work are found in Molden and Sakthivadivel (1999).
[7]   For example, by writing [Basin welfare/Available water] = [Basin welfare/Total depletion] × [Total depletion/Total diversion] × [Total diversion/Water available], we can signify alternative routes to water productivity.

In water-scarce tropical countries with high population density everywhere, as in South Asia and China, maximizing basin welfare involves working on both the components. Increasing the productivity of diverted water is certainly important; but equally important is the need to maximize the proportion of precipitation and inflows into a basin that can be diverted before they are lost to non-beneficial depletion.

It is against this backdrop that we need to consider the growing mass movement for rainwater harvesting and groundwater recharge in western India (Shah, 2000). The region has amongst the highest wind speeds encountered anywhere in the world, it has high mean temperatures for 9 months, rainfall varies between 300 and 800 mm/year, and population density is 300–500 persons per km^2 in the catchment areas as well as in the downstream areas. The greatest challenge for rural communities is surviving the annual pre-monsoon drought in April and May, which is made infinitely more daunting by regular failure of monsoon rains. During the pre-monsoon months, leave-alone growing crops, ensuring adequate drinking water for humans and cattle is the great challenge, especially in the catchment areas of river basins. While government investment programmes concentrated on building large reservoirs downstream to support irrigation and municipal water supplies to towns, the problems of the people living in the catchment areas remained unaddressed.

Disenchanted with government and public systems, non-governmental organizations (NGOs) and communities are finding their own solutions. The past decade has witnessed a massive popular awakening as the result of the efforts of NGOs like Tarun Bharat Sangh, Pradan, and religious organizations such as the Swadhyaya Pariwar. This has taken the form of rainwater conservation and groundwater recharge work on a scale that governments or public agencies would not be able to manage. The basic motivation that has been driving the movement is to ensure availability of domestic water supply for 2 months before the monsoon and for one or two crop-saving waterings

from wells; there are indications that the movement may well meet this challenge.

Government agencies and scientists (hydrologists in particular) have been dubious about this mass movement. Their argument is that rainwater-harvesting structures upstream merely transfer water, reducing the input into the reservoirs downstream, thereby reducing their productivity. However, this argument' does not resonate with the communities, especially in the upstream areas, which fail to see why they cannot meet their basic water domestic needs instead of feeding reservoirs to irrigate relatively small areas of paddy or cotton. In defence of this popular movement, the Delhi-based Centre for Science and Environment has asked: what does India need more – irrigation or drought proofing? In reply, it has suggested that by a total rethink on 'appropriate' RBM, India can trade drought proofing over vast areas by sacrificing irrigation of small areas.

It has also adduced evidence to show that diverting rainwater in a large number of small water-harvesting structures in a catchment captures and stores more of the scarce precipitation, closer to the communities, than having a large reservoir downstream (Agarwal, 2000). This is because water collected over larger watersheds will have to flow over a larger area before it is collected and a large part will be lost in small puddles and depressions, as soil moisture and evaporation.

For instance, Agarwal cites evidence from the Negev desert in Israel to show that 3000 micro-catchments of 0.1 ha capture five times more water than a single catchment of 300 ha, and this multiple increases in a drought period (p. 9). He also cites results by Michael Evanari, an Israeli scientist that show that 'While a 1 ha watershed in the Negev yielded as much as 95 m^3 of water per ha per year, a 345 ha watershed yielded only 24 m^3/ha/year. In other words, as much as 75% of the water that could be collected was lost. This loss was even higher in a drought year.' Agarwal cites Evanari: '. . . during drought years with less than 50 mm of rainfall, watersheds larger than 50 ha will not produce any appreciable

water yield while small natural watersheds will yield 20–40 m³/ha, and micro-
catchments (smaller than 0.1 ha) as much as
80–100 m³/ha.'

Before irrigated crop production,
semiarid India needs drinking water for its
dispersed rural population during the 9
months without rainfall. Many Indian
observers think that the answer is not piped
water supply schemes but decentralized
rainwater harvesting.[8]

3.5 Organization of the Water Sector

Developed-country water sectors, which
have evolved over decades of public intervention, tend to be highly organized and
formalized, with most of the water delivered – and most of the users served – by
service providers in the organized sector.
In low-income countries, a vast majority of
water users – the poorest ones – get their
water directly from rain and from local
private or community storage without any
significant mediation from public agencies
or organized service providers. The notion
of professional water service providers is
alien to a majority of rural South Asians and
Africans.

As a society evolves and its economy,
as well as its water sector, matures, the bulk
of the water delivered to ultimate users
is produced, developed, planned and allocated – in general, managed – by formal
organizations, businesses or utilities. In
Israel, for example, 70% of the water supply
in the country is managed by Mekorot, a
State-owned water company that operates
the National Water Carrier, the pipeline
system that moves water from Lake Galilee
to the Negev Desert, and is in urban water
retail, desalination and sewerage treatment
businesses (Saleth and Dinar, 2000, p. 185).
When the bulk of the users and uses are
served through the formal sector, resource

governance becomes feasible, even simple.
If a basin management regime wants to
increase the water price to domestic users
by 5%, or make a law intended to change
the way business is done, it can do so with
the confidence that it will stick. However,
this is not true when the bulk of the water
users and uses are served by an informal
sector where service providers are not even
registered.

In comparing the Australian success
with containing agricultural pollution of
water with the Chinese situation, Hu (1999,
p. 327) laments that while the small number
of large Australian farmers are served by a
range of local organizations, such as sugar,
rice and cattle associations, which serve as
vehicles not only for new knowledge and
technical advice but also for implementing
new rules and laws, in China, 'given the
small scale of farming units and the large
number of farmers, it is difficult to control
chemical and pesticide application, removal
of vegetation, erosion and water resource
exploitation.' In South Africa, over 90% of
water is managed by formal organizations
but 90% of rural people, the black irrigators
in former homelands, are almost wholly in
the informal sector, far out of the reach of the
organized systems.

Ignoring the complexity of dealing with
the informal water sectors in the developing
world can lead to misleading analyses. In
the perspective of Saleth and Dinar (2000,
p. 186), for example, the institutional reform
challenge in South Africa 'lies in translating
the provision of its water law and water
policy without creating much uncertainty
among private investors.' In our view, these
are easily done. The real challenge with
which the government of South Africa is
struggling is in reaching the reform to the
black communities in the former homelands, who operate in the informal water
sector. As hard as the government is trying,
this is not proving to be easy. About the
process of Catchment Management Agency

---

[8]   The Centre for Science and Environment has estimated the average area needed per village to capture
sufficient water to meet every household's drinking and cooking water requirement in various regions with
varying climate, precipitation and demographic conditions. The average for India as a whole was all of 1.14 ha
per village in a normal year and 2.28 ha per village in a drought year.

(CMA) formation in Olifants, South Africa, Merrey (2000, p. 9) writes:

> rural communities were unaware of the provisions of the new water law and the CMA process, despite the efforts to inform people and offer them opportunities to express their views. Small-scale farmers had not heard about the CMA . . . [But] the Irrigation Boards providing water to large commercial farmers were participating actively in the process . . .

A small number of large stakeholders is easy to work with. The ballgame changes fundamentally once we have to deal with a huge number of tiny stakeholders.

One way the informal sector can be 'formalized' is through grassroots user organizations, and the global irrigation management transfer (IMT) initiatives to organize irrigators into water users associations is partly motivated by the need to bring them into the formal sector. In this too, small numbers of large users in the developed world have an advantage over large numbers of small users in the developing world. All manner of user associations form spontaneously in countries like the USA and Australia. These institutional models are constantly being tried out in developing countries but, here, these generally break down when faced with large numbers of small stakeholders who face such diverse constraints in their livelihood systems that they are at best apathetic towards each of them.

Thus, for example, IMT to water user organizations has unambiguously succeeded in the USA, New Zealand, Colombia, Turkey and Mexico, all situations of medium to large commercial or export farmers who run their farms as wealth-creating enterprises. In contrast, nowhere in low-income Asia, barring a few 'islands of excellence', including the much-researched Philippines, has IMT held out the promise of long-term sustainability. White commercial farmers in South Africa took to Irrigation Boards like ducks to water. In African smallholder black irrigation schemes, there seems little chance that IMT will take off at all unless it is preceded by a wide-ranging

intervention to make smallholder farming itself viable (e.g. Shah *et al.*, 2000).

One standard refrain of institutional discussions in the water sector is get water law and get it 'right'. It is often the case, however, that the problem is not passing a law but in enforcing it in a society with a large number of tiny stakeholders operating in the informal sector with little or no linkage with meso- and macro-level resource governance structures. This is why many governments in Asia readily pass Acts but spend years before converting these into implementable laws and regulations.

There are also cases of countries that have passed laws which have come totally unstuck. Sri Lanka has been debating a water law – which has 'all the right ingredients' (Saleth and Dinar, 2000) – since the early 1980s, but is yet to enact it. This is presumably because it is difficult to figure how to make all the 'right ingredients' – water permit systems, full-cost pricing, water courts and explicit water policy statement – actually work in ways that make a significant difference to the management of water resources in a country where 50–70% of the rural people acquire their water not through water supply service utilities/companies but straight from nature or from local storage in small community tanks.

India adopted a water policy in 1987, but nothing changed as a consequence, and it is now working on a new one. Many Indian states have likewise been debating groundwater laws for 30 years and a dozen or so drafts are in circulation. The legislative assembly of Gujarat, a state with severe groundwater overdraft problems, passed a bill as far back as in 1974, but the Chief Minister refused to make it into a law. His reasons were convincing. First, he was unable to see how the law could be effectively enforced on a million small private pumpers scattered throughout a huge countryside, and second, he was certain that it would become one more instrument of rent-seeking for the local bureaucracy (Shah, 1993).

'Get the price right' is another old prescription to make water an economic good.

Now that water scarcity in many parts of the world is real, it would be naïve to question the value of pricing, not so much for revenue collection but to signal the scarcity value of water to users. There can be no serious debate on whether the view of water as a 'scarce but free' resource is tenable in today's context. The real issue is making the price of water stick in a situation where a majority of users are in the informal sector and do not go to anyone except the rain-gods for getting their water.

Even in canal irrigation systems in South Asia, which are in the formal sector, many political leaders and senior administrators would become open to volumetric pricing of water to promote efficient use, if only the logistics of doing so were simple and cost-effective, what with the large number of small irrigators in the commands of Asian systems. After all, paying high prices for high-quality irrigation service is common for millions of resource-poor buyers of pump irrigation in India, Pakistan, Bangladesh and Nepal. Most people would avoid paying the full-cost price if not paying were an option, as is the case in many developing-country water sectors.

The high transaction cost of monitoring water use and collecting water charge is the central issue in water pricing, rather than the politicians' propensity towards giving away largesse. This will soon be evident in South Africa, where a new water pricing policy will be easy to enforce on large commercial farmers, since transaction cost of monitoring and collection will be low, but very difficult to apply in areas of black irrigation, dominated as they are by large numbers of small users.

Developed-country institutions have not solved the problem of serving or regulating large numbers of small users. Indeed, they have not yet found satisfactory ways of dealing with moderate numbers of large users. In New South Wales, Queensland and Victoria, the existing law confers on every occupier of land the right to take and use water for domestic consumptive purposes, watering stock, irrigating home gardens and non-commercial crops on a maximum of 2 ha (MacDonald and Young, 2000, p. 24).

If this exemption were applied to India, it would cover over 80% of all land and over 90% of all people. In South Africa, it would cover 90% of all users though only 10% of land and water. In South Asia, South-East Asia and North China, groundwater is the most valuable and threatened resource, and protecting groundwater from over-development is probably among the top three priorities in this region. Yet doing so is proving to be a challenge precisely because groundwater is in the informal sector.

In the question of how best to deal with South Asia's 20 million tubewell owners in the informal sector, the experiences of Murray–Darling or Mississippi do not have many practical lessons to offer. Even in 'highly evolved' river basins, sustainable management of groundwater is at best problematic, and at worst, as hopeless as in India and Pakistan. Murray–Darling has tried groundwater regulation but it is not certain if it has worked. Access to groundwater in New South Wales is regulated by licences under the Water Administration Act of 1986. However,

> over much of New South Wales, undeveloped licences were not cancelled. In retrospect, this has proved an administrative disaster as, in a number of areas, the total volume of licences issued is well in excess of estimated sustained yield. (MacDonald and Young, 2000, p. 23)

In California's Central Valley, groundwater over-exploitation is a 60-year-old problem, yet in his case study of basin management, Svendsen (2000) concludes that:

> Groundwater is the most lightly planned and regulated segment of the state's water resources. There is little control over abstractions and, on average, the state is in a serious overdraft situation.

Even in middle-income countries, where major institutional reforms have been initiated in recent years, groundwater over-exploitation has defied solution. Spain, one of the European countries that suffer agricultural over-exploitation of groundwater, has instituted sweeping reforms that will affect surface water but have little to do with groundwater (Saleth and Dinar, 2000).

Mexico's aquifers too are amongst the most over-developed. IWMI researchers based in Guanajuato state, one of Mexico's agriculturally dynamic regions, found water tables in ten aquifers they studied declining at average annual rates of 1.79–3.3 m/year during recent years (Wester *et al.*, 1999, p. 9). An institutional solution is being tried here through the establishment of Aquifer Management Councils called COTAS (Consejos Técnicos de Aguas). IWMI researchers in Guanajuato are, however, sceptical: '. . . several factors bode ill for their (COTAS') future effectiveness in arresting groundwater depletion . . .'.

Finally, for top echelons of national decision makers, it is always easy to take hard decisions, which do not affect a large proportion of a nation's population in a seriously adverse manner. Political leaders and water-sector leaders in emerging economies constantly face pressures to be myopic and adopt postures that are at odds with the ideal of IRBM. The most powerful and compelling pressures emerge from their own internal social realities. In low-income agrarian societies like in South Asia and much of Africa, food security and poverty alleviation will continue to remain prime concerns for decades to come.

When several poor states are involved in a basin – such as India, Nepal and Bangladesh in the Ganga–Meghana–Brahmaputra basin, or the Central Asian states in the Aral Sea – coordinating mechanisms tend to operate at suboptimal levels because national leaders are under pressure to maximize their national interests. It has been argued that the Aral Sea crisis is the outcome of the compelling need of the political leaders in the Central Asian states to ensure food security as well as water-intensive cotton cultivation for export, both at once. A major move to reverse the desiccation of the Aral Sea, the Amu Darya and the Syr Darya will have to wait until something changes the dominant perception of the political leadership in Turkmenistan and Uzbekistan that cessation of cotton monoculture will have politically and socially destabilizing consequences.

3.6 Stage of Socioeconomic Development

What factors might influence the pace of institutional change in developing-country water sectors? Saleth and Dinar (2000) suggest that as water scarcity intensifies, opportunity costs imposed by missing or malfunctioning institutions will increase and transaction costs of institutional change will decline, which together will determine the pace of institutional change in developing countries. A competing hypothesis is offered by the application of Kuznets' curve to natural resource management by societies. Recently, there have been attempts to fit an environmental Kuznets curve to deforestation using cross-country data (Bhattarai and Hammig, 2000).

The environmental Kuznets curve (EKC) poses an inverted U relationship between economic growth and environmental degradation (Fig. 3.4). The core hypothesis is that, as economies grow, they use natural resources as a factor of wealth creation, but as per capita real income grows, demand for environmental amenity grows and there is greater demand and support for environmental protection.

Although the empirical results of some of this econometric work are far from conclusive, intuitively, it seems compelling to suppose that the income elasticity of demand for environmental amenities is lower at low per capita incomes (as in Bangladesh and Burkina Faso) than at high per capita incomes. It follows, therefore, that highly evolved economies of the Western world would have greater demand, capacity and collective will to fix the environmental problems resulting from natural resource mismanagement than low-income emerging economies. In many Western countries, where per capita income growth to present levels took 200 years or more, the EKC effect also took centuries to work out. Historical evidence suggests deforestation in Europe was at its peak at the time of the Industrial Revolution, and the area under forests began to increase long after economic prosperity ensued (Bhattarai and Hammig, 2000).

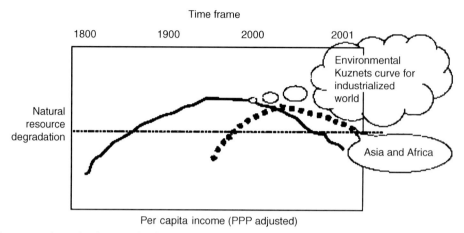

Fig. 3.4.   Relationship between level of economic growth and natural resource degradation.

Much the same relationship seems to hold in the case of water resource management. Countries with highly developed water institutions are also those that have evolved industrially. In contrast, it is difficult to find a low-income agrarian society that has highly developed water institutions. Interestingly, some sketchy evidence suggests that the period of decline followed by upswing gets telescoped in economies like Japan and Taiwan that have grown their industrial output and employment rapidly over a relatively short period.

In Table 3.1, we present the data set for 57 countries organized around their per capita water and arable land availability. The figures alongside the country names are their respective per capita CO_2 emissions, which is one of the best correlates of GDP per capita as well as the Human Development Index. Mean per capita availability of water and arable land, along with CO_2 emission, are used to divide the countries into eight categories.

Countries in categories B1, C1 and D1 are poor in water and/or arable land resources; but these are rapidly becoming post-agrarian societies where pressure on water and land from irrigated farming will rapidly ease. The social and economic costs of fixing water mismanagement in these countries already are or will soon be within acceptable limits.

It is notable that A1 represents the category of countries from which most models of

effective water institutions emerge, and these are offered to countries in D2 category which have the least water, land and CO_2 emissions. Countries in A1 are amongst the best endowed with both water and land. As a result, despite being highly industrialized (as indicated by their high CO_2 emissions), these still have large, wealth-creating agriculture and agroindustry sectors that absorb a very small proportion of their populations.

In the D2 category, poor land and water resource endowments combine with high population pressure, but ironically, their most critical problem is their low CO_2 emission. Industrial growth, urbanization and transfer of people from agricultural to off-farm livelihoods seems the only way pressure on land and water will ease. Many of these countries will, over the coming decades, more likely take the Kuznets-curve route that Japan and Taiwan took than the one that Australia and the USA took.

In Taiwan, where rapid industrial growth and urbanization have resulted in a 40% decline in irrigated areas over recent decades, the popular outlook towards water management issues has undergone a fundamental transformation. Over 90% of Taiwan's irrigators have become part-time farmers, and income from industrial employment far outweighs agricultural incomes. There have been major increases in demand for environmental amenities and the touristic value of former irrigation

Table 3.1.   Natural resource availability and economic growth.

	Per capita water > mean (10,460 m³); values of CO_2 emissions/capita, countries classified as high (1) or low (2)				Per capita water < mean (10,460 m³); values of CO_2 emissions/capita, countries classified as high (1) or low (2)			
Per capita arable land > mean (0.37 ha)	A1		A2		B1		B2	
	Australia	4.61	Argentina	1.01	Denmark	2.95	Sudan	0.1
	Canada	3.73	Brazil	0.46	Greece	2.09	Syria	0.83
	USA	5.37			Kazakhstan	2.89	Zambia	0.08
	New Zealand	2.18			Libya	2.18	Turkey	0.78
	Hungary	1.59			Spain	1.6	Afghanistan	0.02
Per capita arable land < mean (0.37 ha)	C1		C2		D1		D2	
	Malaysia	1.58	Cambodia	0.01	Switzerland	0.68	Zimbabwe	0.45
	Sweden	1.67	Chile	0.92	South Africa	2.1	Tanzania	0.02
	Vietnam	1.77	Indonesia	0.33	Singapore	5.32	Sri Lanka	0.11
			Laos	0.02	Saudi Arabia	3.88	Senegal	0.11
			Mexico	1.02	Oman	1.85	Philippines	0.25
					Japan	2.53	Peru	0.3
					Italy	1.92	Pakistan	0.18
					N. Korea	2.46	Nepal	0.02
					S. Korea	3.07	Kenya	0.07
					Israel	2.5	India	0.29
					Belgium	2.86	Ethiopia	0.02
					France	1.69	Egypt	0.42
							China	0.74
							Botswana	0.37
							Bangladesh	0.05
							Iraq	1.21
							Iran	1.15
							Syria	0.83

Source: Engelman *et al.* (2000).

structures. All these have resulted in substantial private investments in improving water quality and aquatic ecology. Taiwan has amongst the highest population densities we find anywhere in the world, yet its water institutions will soon approach those in high-income Western countries rather than low-income Asian countries, which share high population density with Taiwan.

The Kuznets curve hypothesis looks at the relationship only from the angle of demand for environmental amenities, but there is also the supply side to it. More and higher-quality resources are applied to natural resource management in high-income countries than in low-income countries. Consider the budget of the water departments. The California State Department of Water Resources has 2000 employees, mostly professionals, who operate an annual budget of US$1 billion (Svendsen, 2000). The Gujarat Department of Water Resources probably employs as many engineers but operates a budget of less than US$10 million. The upshot of this discussion is that, over a decadal time frame, economic growth is probably both the cause of and the needed response to the problem of natural resource mismanagement, and, if the experience of Japan and Taiwan is any guide, the period over which the interaction between the two plays out need not run into centuries as it did in the case of Europe, but it can be telescoped from centuries to decades.

3.7 Conclusion

In this chapter, we have made an attempt to explore why efforts to transfer the institutional models of RBM from developed countries to developing ones have not met with the desired success. The contexts in which reforms are tried in developing countries are vastly different – in their hydrologic and climatic conditions, in their demographics, in their socioeconomic conditions and in the way their water sectors are currently organized – from the context of the countries in which the models first succeeded. Successful institutional reforms in the water sector worldwide have tended to have common over-arching patterns – they have focused largely on management of surface water bodies, they have aimed at improving the productivity of publicly diverted large water bodies, they have largely ignored groundwater, and have not had to contend with sizeable informal water sectors, and they have focused on blue-water productivity and largely ignored green water.

The problems that successful institutional models have resolved – poor water quality, degraded wet lands, maintaining navigation and dealing with occasional floods – are often not of paramount interest in developing-country contexts. The problems that developing countries find critical and insurmountable either have remained unresolved in developed-country river basins, e.g. groundwater over-exploitation, or are rendered irrelevant by their evolutionary process, as with using irrigation as a means to provide poor people with livelihoods and food security. This does not by any means imply that developed-country experience has no lessons to offer to the developing world. Drawing such a conclusion would be naïve in the extreme. What it does mean, however, is that imposing institutional models uncritically in vastly different socio-ecological contexts can be dysfunctional and even counter-productive.

What it also means is that we need to take a broader view of institutional change. An extraordinary aspect of the institutional discussion in the global water sector is how very narrowly it has focused on things that governments can do: make laws, set up regulatory organizations, turn over irrigation systems, specify property rights. A recent review of institutional changes in the global water sector in 11 countries by Saleth and Dinar (2000), for example, treats water law, water policy and water administration as the three pillars of institutional analysis. This makes water purely the government's business, quite contrary to the slogan popularized by the World Water Council to 'make water everyone's business'. If institutional change is about how societies adapt to new demands, its study has to deal with more than what just the governments do. People, businesses, exchange institutions, civil society institutions, religions and movements – all these must be covered in the ambit of institutional analysis (e.g. Mestre, 1997, cited in Merrey, 2000, p. 5).

Which elements of the Murray–Darling experience can be sensibly applied in which developing-country context is certainly an important and interesting analytical enterprise, but equally, or even more important is the need to listen to voices from the grass-roots. If people living, for example, in the Deduru Oya basin in Sri Lanka are facing water scarcity, they are going to begin to do something about it. Likewise, if the government of South Africa withdraws from the management of smallholder irrigation schemes in the Olifants basin, the smallholders will soon respond in some way. What institutional reform makes best sense in Deduru Oya or Olifants should best emerge from understanding the respective realities of these basins. A broad understanding of what has worked elsewhere including in the developed world might offer a good backdrop to the design of institutional interventions, but it might be unrealistic to expect much more. Copycat institutional reform would be outright disastrous.

In understanding how societies adapt their institutions to changing demands, Nobel Laureate Oliver Williamson (1999) suggests the criticality of four levels of social analysis as outlined in Fig. 3.5. The top level is referred to as the social embeddedness level where customs, traditions, mores and

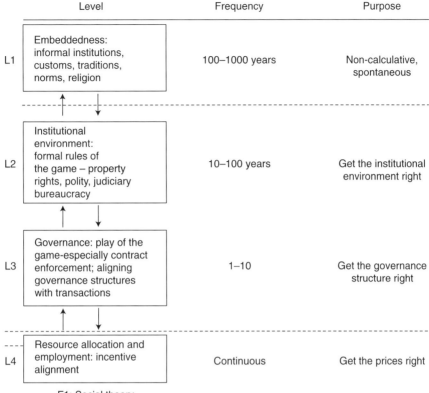

Level	Frequency	Purpose
L1 Embeddedness: informal institutions, customs, traditions, norms, religion	100–1000 years	Non-calculative, spontaneous
L2 Institutional environment: formal rules of the game – property rights, polity, judiciary bureaucracy	10–100 years	Get the institutional environment right
L3 Governance: play of the game-especially contract enforcement; aligning governance structures with transactions	1–10	Get the governance structure right
L4 Resource allocation and employment: incentive alignment	Continuous	Get the prices right

F1: Social theory
F2: Economics of property rights and positive political theory
F3: Transaction cost economics
F4: Neo-classical economics, principal/agent theory

Source: Oliver E Willamson. 1999. The new Institutional Economics: Taking Stock/Looking Ahead , Business and Public Policy Working paper BPP-76, University of California, Berkeley.

Fig. 3.5. Four levels of institutional change that explain how societies adapt to new demands.

religion are located. Institutions at this level change very slowly because of the spontaneous origin of these practices in which 'deliberative choice of a calculative kind is minimally implicated'. At the second level – where the institutional environment of a society is involved – evolutionary processes play a big role; but opportunities for design present themselves through formal rules, constitutions, laws, property rights. The challenge here is getting the rules of the game right. The definition and enforcement of property rights and contract law are critical features here. Also critical is understanding how things actually work, 'warts and all', in some settings but not in others.

However, it is one thing to get the rules of the game (institutional environment) right. It is quite another to get the play of the game (enforcement of contracts/property rights) right. Which leads to the third level of institutional analysis: transaction costs of enforcement of contracts and property rights, and the governance structures through which this is done. Governance – through markets, hybrids, firms, bureaus – is an effort to craft order, thereby to mitigate conflict and realize mutual gains; good governance structures craft order by reshaping incentives, which leads to the fourth level of social analysis – getting the incentives right.

Discussion of water policy and institutions in the developing-country context has focused a great deal on levels 2, 3 and 4 and little on level 1. More, it has tended to underplay the interactions between levels. Many populous developing countries will feel a lot wiser about IRBM if we learn more about how level 1 operates in their respective contexts and how the interaction between 2 and 3, and 3 and 4 can work better. How to create property rights that affect users' behaviour is more important than exhortations that clear property rights be created. Understanding how to enforce a groundwater law meaningfully on 20 million private pumpers scattered throughout the South Asian countryside is more helpful than pushing a groundwater law. How to monitor water use and collect canal irrigation charges cost-effectively is more in order than discussing whether irrigation subsidies should be eliminated.

Acknowledgements

This work is supported by a Research Grant from BMZ (Germany's Ministry of Economic Co-operation) to IWMI to study Institutional Support Systems for Water Management in Water-stressed River Basins.

The authors gratefully acknowledge the contributions of colleagues in developing these ideas: Matsuno for supplying literature on water management institutions in Japan; Hilmy Sally for help with an Excel chart; David Molden for discussions on water accounting; Randy Barker for drawing attention to Williamson's excellent article; Madhusudan Bhattarai, from whose work the ideas on the environmental Kuznets curve are drawn; Frank Rijsberman for a stimulating discussion in a car ride in Hyderabad on what are the meaningful issues in the debate on water pricing; Tissa Bandaragoda and M. Samad for many interesting discussions on basin management research; and Doug Merrey for introducing the author to the field of IRBM. The authors own the responsibility for the argument, with all its limitations and flaws.

References

Agarwal, A. (2000) *Drought? Try Capturing the Rain*. Centre for Science and Environment, Delhi.

Bhattarai, M. and Hammig, M. (2000) Institutions and the environmental Kuznets Curve for deforestation: a cross-country analysis for Latin America, Africa and Asia. *World Development*, forthcoming.

Birch, A. and Taylor, P. (1999) International mentoring; application of Australian experience for Sri Lankan water sector reforms under Technical Assistance of the Asian Development Bank. *Water International* 24, 329–340.

Engelman, R., Richard, P., Bonnie Dye, C., Gardner-Outlaw, T. and Wisenewski, J. (2000) *People in the Balance: Population and Natural Resources at the Turn of the Millennium*. Population Action International, Washington.

Global Water Partnership (2000) *Integrated Water Resources Management*. Technical Advisory Committee of GWP, Stockholm.

Hu, X. (1999) Integrated catchment management in China: application of the Australian experience. *Water International* 24, 323–328.

Hunt, C. (1999) Transposing of water policies from developed to developing countries: the case of user pays. *Water International* 24, 293–306.

Jacobs, J.W. (1999) Comparing river basin development experiences in the Mississippi and the Mekong. *Water International* 24, 196–203.

MacDonald, D.H. and Young, M. (2000) A case study of the Murray–Darling Basin: draft preliminary report. Prepared under commission for IWMI. CSIRO, Sydney.

Malano, H., Bryant, M. and Turral, H. (1999) Management of water resources: can Australian experiences be transferred to Vietnam? *Water International* 24, 307–315.

Merrey, D. (2000) Creating institutional arrangements for managing water-scarce river basins: emerging research results. Paper presented at the session on 'Enough Water for All', at the Global Dialogue on the Role of the Village in 21st Century: Crops, Jobs, Livelihoods, 15–17 August, Hanover.

Molden, D. and Sakthivadivel, R. (1999) Water accounting to assess uses and productivity of water. *Water Resources Development* 155, 55–71.

Mostert, E., van Beek, E., Bouman, N.W.M., Hey, E., Savenije, H.H.G. and Thissen, W.A.H.

(1999) River basin management and planning. Keynote paper for International Workshop on River Basin Management, The Hague, 27–29 October, 1999.

Saleth, M. and Dinar, A. (2000) Institutional changes in global water sector: trends, patterns and implications. *Water Policy* 2, 175–199.

Shah, T. (1993) *Water Markets and Irrigation Development: Political Economy and Practical Policy.* Oxford University Press, Bombay.

Shah, T. (2000) Mobilizing social energy against environmental challenge: understanding the groundwater recharge movement in Western India. Natural Resource Forum, August.

Shah, T., van Koppen, B., Merrey, D., de Lange, M. and Samad, M. (2000) Institutional alternatives in African small holder agriculture: lessons from international experience in irrigation management transfer. Paper presented at National Policy Workshop on Irrigation Management Transfer and Rehabilitation of Small Holder Irrigation Schemes, 21 June, Hazy View, Mpumalanga, South Africa.

Sutcliffe, J.V. (1995) Hydrology in the developing world. In: G.W. Kite (ed.) *Time and the River: Essays by Eminent Hydrologists.* Water Resources Publications, Saskatoon.

Svendsen, M. (2000) Basin management in a mature closed basin: the case of California's Central Valley. In: C.L. Abernethy (ed.) *Intersectoral Management of River Basins.* IWMI, Colombo, Sri Lanka and DSE, Feldafing, Germany.

Wester, P., Pimentel, B.M. and Scott, C. (1999) Institutional responses to groundwater depletion: the aquifer management councils in the State of Guanajuato. Paper presented at the International Symposium on Integrated Water Management in Agriculture, Gomez Palacio, Mexico, 16–18 June.

Williamson, O.E. (1999) The new institutional economics: taking stock/looking ahead. Business and public policy. Working Paper BPP-76, University of California, Berkeley.

4 Making Sound Decisions: Information Needs for Basin Water Management

Martin Burton and David Molden

4.1 Introduction

This chapter presents an analysis of information needs in water-scarce river basins from two perspectives – those of the basin manager and of the manager of a locally controlled irrigation system (LCIS). Information is central to planning and managing the water resource within a river basin. Information requirements develop and grow over time as pressure on the water resource increases with demographic and economic changes (see Chapters 1 and 2). Information plays a crucial role in quantifying the abundance and quality of the available resource and the demands placed on it, and is central to decision making on water allocation at both scheme and river basin level, and to formulation of policies and strategies for river basin development.

As water becomes scarcer, data and information become increasingly important for irrigation system managers, since the need grows to make the most productive use of available supplies. In addition to looking at the internal allocation and management of the available water, managers of irrigation systems must also look externally to ensure that their systems acquire and retain their share of the available resource. As irrigation is often the largest consumptive user of water in a river basin, irrigation managers must often justify the use, efficiency and productivity of water in competition and comparison with other uses and users. In

this context, data and information play a central role.

4.2 Use and Users of Information

There is a wide variety of uses and users of information related to water resources within a river basin (Table 4.1). The river basin can be divided into water uses related to watershed land use, in-stream water use, extractive water use and environmental water use. The in-stream uses are those wherein the water is used without being removed from the river, such as hydropower or recreation. They generally result in minimal depletion of available water. Extractive uses such as irrigation, on the other hand, involve removing water from a river or aquifer, and generally result in larger depletion of the resource from the basin through evapotranspiration. Removal of water from a river or aquifer is not absolute, however. For example, most of the water used for domestic and industrial purposes is available for reuse, either after treatment or through dilution in fresh water.

Information uses are many and varied, and can be categorized into development and master planning, operational management, water sharing and allocation, and research. Data collected for one purpose, such as operational management, may be of use for other purposes, such as development

Table 4.1. Typical water uses, information uses and users within a river basin.

Water uses	
Watershed water uses	• Lakes/reservoirs
	• Forests
	• Natural vegetation
In-stream water uses	• Hydropower
	• Recreation
	• Navigation
	• Fisheries
Extractive water uses	• Irrigation (surface/groundwater)
	• Potable water (surface/groundwater)
	• Industrial water, including mining (surface/groundwater)
Environmental water uses	• Aquatic, wetlands and floodplain environment and ecology
	• Drainage disposal
	• Waste dilution and disposal
	• Repelling salinity intrusions
	• Erosion control
Information uses	
Development and master planning	• Planning and forecasting
	• Decision making in relation to resource development and protection
Water sharing and allocation	• Resource management and allocation
	• Allocation of water rights
	• Rule formulation
	• Pricing
	• Dialogue with, and amongst, users
Operational management	• Flow control and regulation
	• Flood control, protection and warning
	• Effluent control
	• Monitoring and evaluation (abstractions, effluent levels, environment, etc.)
	• Infrastructure asset management
	• Conflict resolution
Research	• Water resources, irrigation, environment, ecology, etc.
Information users	
Government	• Ministries of: Water Resources, Irrigation, Agriculture and Livestock, Energy, Hydrology and Meteorology, Health, Environment and Natural Resources, Fisheries, Forestry, Navigation and Marine Transport, Planning and Development
	• Legislatures
	• State, regional or local government
	• Municipalities
Regulatory and management authorities	• River boards, river basin councils, drainage boards
	• Regulatory bodies (rivers, groundwater, environment, etc.)
	• Courts
Companies, groups and associations	• Industry (manufacturing, services, mines, forestry, etc.)
	• Associations (irrigation, rural water supply, environmental lobbies, etc.)
	• Universities, research centres and training centres
	• Development agencies and agents
	• Non-governmental organizations
Individuals	• Domestic household users
	• Irrigation farmers
	• Livestock owners
	• Recreators

and master planning. Often, however, requirements in different applications differ enough that data collected for one purpose are not usable in another.

Information users can be subdivided into government ministries and agencies, authorities, groups of individuals (such as in a company or an association) and individuals. The relationships that exist with a given river basin between government agencies, groups and individuals are a function of the institutional framework that has evolved over time. The nature of these relationships governs the manner in which data are shared and used.

Data needs evolve over time to suit the changing needs of a developing basin. The drivers for change (population growth, agricultural and industrial development, concern for the environment, increasing democratization, etc.) lead to changes in water use, water quality and institutions, which in turn govern the type and extent of data collected, processed, analysed and disseminated, with new needs building on existing practices. As a democratic society matures, power and decision making are transferred to a wider community. Information processes become more demand driven, with those responsible for collecting and processing data having to become more responsive to the data users. The nature of the transaction between the provider and the data user may be hierarchical, commercial, informal or statutory, or it may not be clearly defined. As time passes, these relationships need to become more clearly and closely defined if the best use is to be made of the available resource.

4.3 Information System Tasks

Information system tasks range from data collection, through processing and analysis, to storage and retention (Fig. 4.1). A further stage, which is becoming increasingly important, and which is relevant to LCISs, is that of disseminating both data and the findings arising from that data. In some countries, this stage has been reached. In others, data are not widely disseminated outside the agency that collects and processes the data. In these latter instances, access to data by third parties is sometimes impaired.

The following sections provide a brief overview of the information system tasks necessary for both river basin management and managing LCISs. A distinction is made between data and information, with data being a series of observations or measurements, and information being the results derived from processing the data. Thus, the same data can be processed in different ways to provide information for different uses or users.

4.3.1 Formulation of an information strategy and management information system

It is essential for managers of any system, be it a river basin or an irrigation scheme, to formulate a *management information strategy* and from that to develop a *management information system*. The information strategy is the pattern or plan that integrates the organization's information needs into a cohesive whole to meet the organization's management requirements. It deals with the information needs, and the resources required for their collection, processing and use. The system is the collection of procedures that implement this strategy. The strategy and associated management information system should address the following questions:

- What are the key management processes?
- What data and information are needed for managing these processes?
- What are the key external factors affecting the enterprise, and what data need to be collected to monitor these over time?
- How, and by whom, will data be collected, processed, analysed and used?

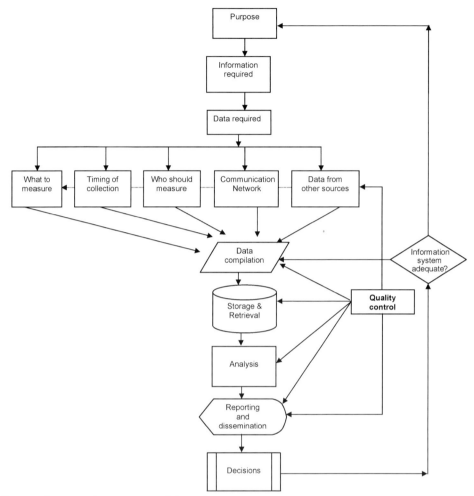

Fig. 4.1.   Structure of a management information system.

- How much will the data collection, processing, analysis and dissemination cost?
- What quality control measures will be required and used?
- How transparent will the information system be?
- To whom will data be disseminated, what data will they require and in what form?

Key management processes in a river basin grow directly from the functions performed. Chapter 1 presented a set of generic river basin management functions and a methodology for analysing their performance, which can be used to guide a detailing of key management processes. Information needs for management will depend on such factors as the size of the basin, its stage of development, the diversity of water users and prevailing institutional arrangements. An important potential external factor is a political demand to transfer water outside the basin or to develop particular uses, such as expanded irrigated agriculture, within the basin. The management information system developed will depend, among other things, on whether there is one river basin management authority, or several separate agencies with responsibility for the management/oversight of the water resource. An information strategy and its associated management information system are dynamic and must change over time to match social and political changes within society (hence the question 'Is the information system adequate?' in Fig. 4.1).

For LCISs, the management processes are relatively straightforward, with data being required in relation to water demands, cropped areas and irrigation fees paid. These data are generally fairly easy to collect and are limited in geographical extent, compared with data for the basin as a whole. External data required relate to rainfall and river flows elsewhere in the river basin, and water abstractions by other users, especially those upstream. Such data may be collected by a government agency or by other irrigation scheme managers. However, gaining access to these data by LCIS managers may be difficult.

Information needs will vary with the stage of development of the water resource, and reflect the level of technology and expertise available. It is crucial that the management information system 'fits together' from start (purpose) to finish (decision making), with systems designed to be congruent with local capabilities and resources at all stages. It is preferable to establish a system that is robust and provides relatively simple, but useful, information, than to establish an 'ideal' one which is over-ambitious and unsustainable. The focus needs to be on the data and information needs of all the actors involved in irrigation system or river basin management, leading to a demand-led information system.

4.3.2 Data collection

Once data needs have been determined during the strategy formulation, attention shifts to data collection. The two key dimensions are *time* and *space*, and data collection need to be distributed throughout both dimensions. Decisions on *where to measure* and the *duration* and *frequency* of measurement should be established when designing the management information system.

Data-collection programmes can be regular, periodic or one-off depending on needs. Regular data collection generally includes river flows, rainfall and water quality measurements. Periodic data collection might include peak or low flow measurements, whilst one-off data collection might relate to a time-bound project where information is required to make decisions for the project. Data-collection procedures can be different in these cases. One feature of a regular data-collection programme is that standard *pro forma* are used to record the data. Different degrees of certainty can be attached to these different types of data. Measured data are often more reliable than reported or estimated data, while observed data may be more dependent on the bias of the observer.

Types of data to be collected comprise the following:

- *Measured* – data obtained from direct measurements, e.g. discharge from depth gauge readings at a measuring structure;
- *Observed* – information used in a qualitative mode based on field observations, such as canal condition, activities of organizations, etc.;
- *Estimated* – data used in a quantitative mode which are derived indirectly by calculation using combinations of measured variables (such as potential evapotranspiration using the Penman–Monteith equation), or which are estimated without direct measurement;
- *Reported* – data obtained from secondary sources, using other peoples' observed, estimated or measured data.

Two major concerns in data collection are *reliability* and *accuracy*. The reliability of the data relates to the degree of assurance that the data are measured and not made up. The accuracy relates to how close measured data are to the 'true' value of the parameter. If data are to be used for management and decision making, it is imperative that the data be reliable. Data collection can be a tedious and time-consuming task for those collecting the data, and checks have to be made to ensure that personnel responsible for these tasks carry them out reliably. It is not uncommon for data to be made up by data collectors to avoid the chore of travelling to a recording site. Accuracy of data collection can be enhanced through trained personnel periodically checking measuring sites (river gauges, rainfall stations, climate stations, etc.) to ensure that they are functioning correctly, and that data are being collected in the correct manner, or by improved measurement technology. Generally, increased accuracy is associated with increased costs.

4.3.3 Storage, retention and retrieval of data

Often overlooked until data are required and found to be lost or missing are procedures for storage, retention and retrieval of data. The management information system should clearly indicate which data should be stored, in what format and for how long. At higher management levels, summaries of the data can be stored, whilst at lower levels the complete data should be retained.

Data storage should be systematic and files clearly labelled. The data may be in the form of sheets of paper, in which case it must be kept in a secure, dry location. Computer data must always be backed up on floppy disk or CD, and the disks kept in a secure location separate from the computer.

Geographical information systems (GISs) are a valuable way of both presenting and storing data, with the significant advantage that the data can be retrieved in different combinations or formats depending on the need.

4.3.4 Data processing and analysis

The procedures for data processing and analysis should be identified as part of the information strategy. It is the use to which the data will be put that drives the data-collection exercise; thus, how the data will be processed and analysed should be established at the outset. For regular data collection of river flows, rainfall, water

quality and the like, data processing and analysis procedures are well established. For other processes, such as performance assessment, procedures are still evolving.

The widespread availability of powerful small computers and software packages has transformed data handling, processing and analysis. Using spreadsheet programs, one is able, relatively simply, to set up procedures for recording, storing, processing and presenting data. Particularly useful is the ability to represent data graphically to show events, trends and relationships. Statistical analysis packages such as SPSS can often access data stored in spreadsheet format and process it to present information on trends and relationships among data.

An example of the value of trend analysis using data collected over several years is discussed in Chapter 8, drawn from study of the Lerma–Chapala river basin, showing the level of Lake Chapala, the receiving body at the tail of this basin. As a consequence of an extensive network of data-collection stations throughout this basin, an important long-term set of data is available for analysis and presentation, leading to better-informed decision making related to the management and development of the basin. The role and importance of these data has increased as the available water resource has become fully utilized.

4.3.5 Reporting and dissemination of data

The reporting and use of data should be identified as part of the information strategy, and tailored to the target audience. For water resources specialists, tables and figures of river discharges and rainfall patterns will be of interest. Politicians, on the other hand, will want summaries and a discussion of the implications of the data, with key issues highlighted.

With the advent of computers and the Internet, wider dissemination of data becomes possible, even with limited budgets. Data can be held on a central computer and accessed through the Internet, with passwords allowing different levels of access to different users. IWMI's World Atlas is an example of such an Internet-based system (http://www.iwmi.cgiar.org/WAtlas/atlas.htm). In addition, the ability to copy data on to CDs, which can then be distributed to interested parties, means that cheap access to data is possible for those that do not have ready access to the Internet.

There is an old adage that knowledge is power, and in many countries, access to data is still restricted, either as a government policy, or as an organizational or individual practice. In some locations, data are collected by government agencies and then sold to other users. This selling of data can be official, with clearly spelt out policies and prices, or unofficial with individuals using their position to gain additional income.

4.3.6 Data quality control

Quality control on water resource data is often undervalued and inadequately implemented. Quality control checks can and should be built into data-collection,

processing, analysis and reporting processes, and personnel responsible for these processes trained in monitoring quality and identifying suspect data. As mentioned previously, supervision of data collection is important, as is explanation to data-collection personnel of the importance of accurate and reliable data, and the use to which the data will be put.

4.4 Basin Stages and Changing Information Needs

In tandem with the development of a river basin, the collection and use of data changes. Figure 4.2 depicts the changing focus for data and information needs as basins mature.

As time passes, quality and regulatory issues come to the fore, and information is required by a wider audience. The data set required for river basin management becomes both more extensive and more detailed, with increased needs for supporting processes, procedures, personnel and

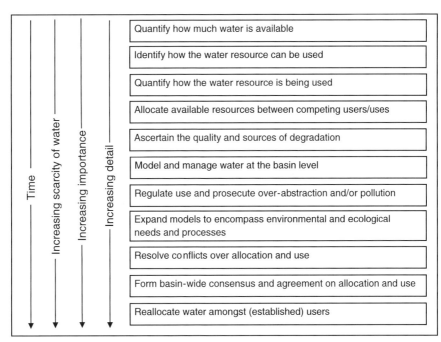

Fig. 4.2. Changing focus for data and information over time within a river basin.

finances to match. As the basin develops, information is collected and used for modelling – initially of the hydrological and hydraulic processes and later for environmental and ecological processes. With increasing water scarcity, conflicts over water use increase, and data are increasingly employed in the resolution of conflicts. With increasing incidence or scale of conflict, there comes a time when steps are required at the basin level to form a consensus on individual, corporate, societal and environmental rights to access and use the water resource, following which there may be a need for reallocation of water amongst uses and users.

As discussed in Chapter 2, three phases can be identified in the development and growth of a river basin: development, utilization and reallocation. Information and data requirements in each phase vary, and are summarized in Table 4.2. The table shows information needs, provides examples of the data required and shows the developments that typically take place in information processes during each phase in response to developments within the river basin.

In the early development stages, much of the data collection is project based, with each project establishing data-collection systems to suit its particular needs. The data-collection needs will follow the needs of the project cycle: planning, design, construction, management, operation and maintenance, and possibly later, rehabilitation. At this stage, there is relatively limited human control over the natural environment, with modest physical infrastructure and localized agreements on flood control and water use.

As time passes, these data systems are integrated by one or another of the government agencies, and the data network becomes more systematized. At this stage, data collection and processing may be driven by established routines and not necessarily by needs. As pressure on the available water resource increases, the need for data grows, and the data system becomes more refined with more data-collection points being established and the data

collection, processing and analysis of the data becoming more sophisticated. Hydrologists develop computer-based water resource models, which, in turn, begin to drive data collection so that the predictive capabilities of the models can be fully utilized.

Over time, the focus shifts from attempting to supply water to match increasing demand to demand management. During the development, utilization and reallocation phases, the approach to data changes, from one in which the data are collected, processed and used by single agencies for their own purposes to a situation where the data are disseminated and made available to a wide range of stakeholders. The power that data ownership holds is shared amongst the various stakeholders to achieve the wider objective of consensus and buy-in to water resources development and management. During the reallocation phase, the focus is on ensuring that the various users limit their abstractions to the quantities licensed, and maintain their effluent discharges within the specified water quality standards.

Potential interventions to address increasing water shortage vary according to the development stage of the river basin. At each stage, the solution revolves around information, as shown in Fig. 4.3. For each problem situation, the nature of the problem is formulated, and data collected and analysed. Solutions can take many forms, involving alternatively, or in combination, construction, legislation, enforcement of existing legislation, improved management or empowerment of organizations or individuals.

4.5 Basin Level Information Needs

The types of basin level information required change both with the phase of development and with the perception of what river basin management comprises. Early guidelines on river basin management tend to emphasize the collection of technical data (climate, river flow, catchment area, land use, etc.), whereas more recent

Table 4.2. Data requirements related to development phases of river basins.

Phase	Data needs	Typical data collected	Developments in information processes
Infancy Localized use only	Rudimentary, limited to water levels and extent of flooding	Flood water levels, flooded areas (through experience)	Demarcation (and avoidance) of flooded areas, correlation of flood extent and flood levels
Development • Water allocation is supply focused • Data collected and used by small number of agencies for specific uses and projects	• Availability of water during the year and extent of agricultural land • Main focus is on surface water, though some interest in groundwater for urban and irrigation development • For initial planning for river basin development	• Project-wise collection of river flow and quality data • Climatic data, particularly rainfall • Land use in riverine plains and extent of agricultural land • Topographic surveys • Aerial photography • Land ownership, traditional/existing water rights	• Initial data collection systems established for individual projects; gradually these are linked up and coordinated by the development agency(s) • Basin-wide hydrometric stations established to gather base data
Utilization • Water allocation is supply focused • Data related processes and procedures well established	• Detailed knowledge of the available water resources, both surface and groundwater, particularly over-year to establish storage patterns for reservoirs and recharge patterns for groundwater. • For river basin master planning	• River flow data throughout the basin • Climatic data throughout the basin • Land ownership and traditional/existing water rights • Groundwater level and quality • Some monitoring of pollution levels	• Data collection procedures standardized and coordinated • Procedures established for monitoring pollution levels • Procedures established for monitoring groundwater depth and quality • Publication of water resources and climatic data • Development of simple water resources models for river basins

continued

Table 4.2. *Continued.*

Phase	Data needs	Typical data collected	Developments in information processes
Re-allocation and restoration • Demand and supply focused • Data related processes and procedures refined and more widely disseminated	• To obtain detailed knowledge of the annual and inter-year water resource situation both for supply and demand • To monitor and control water abstraction by users • To make projections of supply and demand • For water resources modelling, using remote sensing and GIS • For scenario analysis • For river basin master planning • To refine and update supply and demand projections, scenario analysis • To formulate rules for allocation of water during droughts/shortages	• River flow and water quality data throughout the basin • Climatic data throughout the basin • Groundwater level and quality • Pollution levels • Water abstraction by all users • Data for prosecution for over-abstraction and/or pollution • Data analysed from perspective of different water users • Water needs for various environmental processes	• Hydrometric network extended and automated for direct transmission to data collection stations • Groundwater monitoring network extended • Pollution monitoring extended • Further computerization of data collection, processing and analysis • Development of sophisticated water resource models for river basins, with refinement to become an operational tool • Remote sensing incorporated into water management and decision making • Publication of water resources supply and demand information • Analysis and presentation of data for a wider range of stakeholders • Scenario analysis to enable participation in decision making

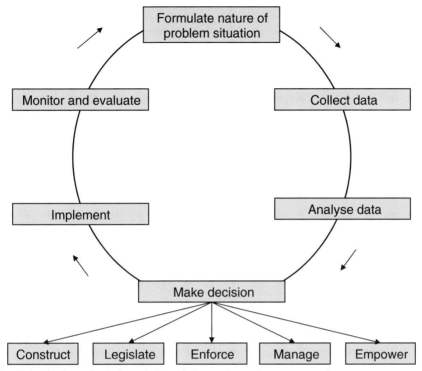

Fig. 4.3.   Role of information in formulating solutions to increasing water scarcity.

publications recommend a broader collection of data encompassing economic, institutional, social and environmental features of the basin and society in general. An example is found in the guidelines developed by the Asian Development Bank following an extended process of consultation and discussion (ADB, 1996). The guidelines list seven key strategies for improving water resources management.

- Prepare and adapt a national water policy and action programme;
- Invest to manage the country's priority river basins;
- Increase the autonomy and accountability of service providers;
- Develop incentives, regulation and awareness for sustainable water use;
- Manage the use of shared water resources and develop cooperation;
- Enhance water information, consultation and partnerships;
- Invest in capacity building, monitoring and learning.

These seven strategies are clearly information and data intensive. A key process is compiling National Water Sector Profiles. The broad content of these profiles is presented in Box 4.1, and encompasses a range of items from general development goals and water resources policy statements, to legal framework, to the physical resource base, demonstrating the holistic approach that is a central feature of integrated water resources management. Whilst the ADB's water sector profiles are country based, many of the profile categories and data requirements are relevant at the river basin level as well.

Some key data requirements for river basin management are presented in Box 4.2. A list of leading indicators for describing river basins is presented in Table 4.3. Collection and presentation of these data allow comparative assessment of different river basins, enabling the relative level of development to be assessed. As can be seen from the data presented in Table 4.3, the three basins are at different stages of development.

Section 1 Country overview
Section 2 National policy environment
2.1 National development goals
2.2 Water resources policy
2.3 Transnational and subnational relations
Section 3 Capacity for water resources management
3.1 Legal base
3.2 Institutional base
3.3 Information base
3.4 Training and human resources development
Section 4 Water resources status
4.1 Water resources and watersheds: the physical base
4.2 Uses of water
4.3 Community values of water
4.4 Supply, demand and sustainability
4.5 Summary: status and trends
Section 5 Financial resources
5.1 Annual expenditure and revenues
5.2 Return on investment
Section 6 Appraisal
6.1 Water sector institutions
6.2 Water resources and watersheds
6.3 Uses of water
6.4 Community values of water
6.5 Sustainability of water resources and use
6.6 Financial performance
6.7 Consistency with external support agency objectives
Section 7 Agenda for action
1 Appendices
2 References
3 Completed and current water-related projects
4 Summary of lessons learnt from completed projects
5 Contact persons
6 Maps of priority river basins
Data Appendices

The East Rapti basin has more than enough water resources to match current needs, the Gediz basin is closing, with pollution a major current issue, whereas the Lerma–Chapala basin is closed and over-exploited.

The heart of any basin management exercise is a water budget for the basin (Molden, 1997). A water budget requires time series data on water entering, being utilized and leaving the basin, as well as information on storage and storage losses.

The budget must include both ground- and surface water. Preparing a basin water budget requires physical data from a network of hydrologic, meteorological and groundwater stations within the basin. These networks often take years to establish, and a number of years of data are required to achieve a minimum level of reliability in the results. In some cases, these networks will have been established through externally funded projects, and finding adequate funds to maintain the networks may have been a problem. Whilst it takes time and effort to establish these systems, particularly the training of staff to collect, process and analyse data, they can deteriorate relatively quickly if funding is inadequate or if sound management is lacking.

Remote sensing is increasingly being used for river basin management. Kite and Droogers (2000) used remotely sensed data to construct an integrated river basin model for the Gediz river basin in western Turkey. For this study, they compiled models at three scales: the field scale, the irrigation scheme scale and the basin scale. Using public domain datasets as well as local area datasets, they constructed these models based on two categories of data – areal and point data. Areal data include topography, land cover, leaf area index (LAI) and soil characteristics. Point data include climate, streamflow and operational rules for dams and regulators. Kite and Droogers point out that two changes in methods of data access and data collection can be observed in recent years. First, data are increasingly available from global datasets, and second, more and more data are collected by remote sensing (RS) instead of conventional ground-based techniques (Fig. 4.4).

For their model, Kite and Droogers collected and used data as summarized in Table 4.4.

Through the modelling, a wide range of information was presented:

- Distribution of land cover;
- Basin-wide soil water-holding capacity;
- Annual basin-wide evapotranspiration;
- Simulated streamflows at specified locations within the basin;

Box 4.2.   Summary of key data for river basin management.

Physical data
- Latitude/longitude
- Catchment area
- River channel length
- River slopes
- Land use types and areas
- Land slopes and areas
- Soil types and areas
- Aquifers (numbers and areas)

Demographic data
- Total population (past, present and projected)
- Population densities
- Population by location (urban/rural)
- Population by work type
- Attainment levels for education (by age and gender)

Institutional
- Development policy
- Water policy
- Water law
- Environmental law
- Land tenure
- Stakeholders – roles and responsibilities
- Water rights

Economic
- National GNP
- Regional or basin GNP
- Average basin per capita GNP

Hydrometric data
- River discharges
- River water levels
- River flood peak discharges
- River base flows
- River sediment load
- River water quality
- Lake/reservoir water levels
- Lake/reservoir volumes
- Lake/reservoir water quality
- Lake/reservoir water temperature

- Lake/reservoir surface evaporation
- Volume of water imported/exported to/from basin

Meteorological and climatic
- Sunshine/radiation hours
- Wind speed
- Air temperature – average/max/min
- Humidity
- Evaporation
- Precipitation
- Precipitation intensity

Groundwater
- Groundwater levels
- Groundwater quality
- Aquifer yields and quality
- Estimate annual groundwater recharge

Agricultural
- Cultivable area
- Irrigable area
- Irrigated area
- Irrigation water abstractions (surface/groundwater)
- Drainage return flows – quantity
- Drainage return flows – quality
- Number of landholders
- Population dependent on irrigated agriculture
- Value of irrigated agricultural production

Potable and wastewater
- Abstraction quantity (surface/groundwater)
- Abstraction quality
- Return flow – quantity
- Return flow – quality
- Number of people supplied

Industrial
- Abstraction quantity (surface/groundwater)

- Abstraction quality
- Return flow – quantity
- Return flow – quality
- Number of people employed

Navigation
- River water levels
- River discharges
- River channels and depths

Hydroelectric power
- Generation capacity
- Discharge requirements and timing
- Maximum discharge requirements and timing
- Minimum discharge requirements and timing

Environmental
- Minimum flow requirements
- Critical flow periods and demands
- Protected areas and water demands
- Required water quality standards

Recreational
- Minimum flow requirements
- Critical flow periods and demands
- Protected areas and water demands
- Required water quality standards

Tourism
- Minimum flow requirements
- Critical flow periods and demands
- Protected areas and water demands
- Required water quality standards

- Seasonal and annual crop transpiration and soil evaporation in irrigated areas;
- Simulated crop yields for actual seasonal climatic conditions in dry and wet years.

The model was used to analyse and assess the current management of the water resource within the basin and to assess the impact, especially on crop production, of possible changes, such as a change in climate.

Table 4.3. Examples of key descriptors and indicators for river basins.

Indicator	East Rapti basin, Nepal	Gediz basin, Turkey	Lerma–Chapala basin, Mexico
Climate classification	Tropical monsoon	Mediterranean	Semiarid to subhumid
Type of basin (open/closing/closed)	Open	Closing	Closed
Catchment area (km²)	3,120	17,700	54,300
Main river channel length (km)	122	275	750
Land use (% by type)	Forest (65%) including National Park; rainfed agriculture (17%); irrigated agriculture (10%); grass and shrub (1.5%); river, streams and sand (5.5%), others (1%)	Maki (30%); forest (26%); rainfed crops (25%); irrigated crops (8%); shrubland (3%); barren (8%)	Rainfed cultivation (37%); irrigated agriculture (20%); scrub (12%); grassland (14%); forest (12%); water bodies (3%); other (2%)
Number of lakes/reservoirs	No constructed dams/reservoirs	2 large; 2 small	27 (large); 1,500 (small)
Population (millions)	0.54	0.67	11
Population density (persons/km²)	166	38	203
Ratio of urban to rural population	25 : 75	60 : 40	70 : 30 (urban defined as settlements with more than 2500 inhabitants)
Annual hydropower generation	Nil	100 GWh	192 GWh
Average annual rainfall (mm)	2008	800 mm (upper catchment) 450 (lower catchment)	712 (av.) 494 (min.) 1,022 (max.)
Average annual evapotranspiration (mm)	987 (ETa)	1,250 (ETp)	1,900
Rainfall months and average amount falling (%)	6 months (93%)	6 months (80%)	6 months (91%)
Dry season months and average amount falling (%)	6 months (7%)	6 months (20%)	6 months (9%)
Average annual river flow volume (MCM)	5,993	940	5,757
Peak flood flow (m³/s)	225	718	1,050
Minimum base flow (m³/s)	35	0	0
Irrigated area (ha)	32,388 (gross)	110,000 (large scale); 25,000 (small)	409,000 (surface); 380,000 (groundwater); 789,000 (total)

Average drop of groundwater level (m/year)	Not falling	Falling	2.1
Annual volume of water use (MCM, %):			
Irrigation (surface/groundwater)	240	660 (surface, 75%); 35 (groundwater, 4%)	6,584 (68%)
Potable water (surface/groundwater)	5.9	134 (14%)	1,351 (14%)[a]
Industrial (surface/groundwater)	0.18	50 (6%)	278 (3%)
Hydropower	Nil	No special releases	39 (included in evaporation from water bodies)
Navigation	None	None	None
Environment	473 (allocation for National Park activities)	4 (current allocation, 1%, more needed)	2,270 (23%) defined as evaporation from water bodies
Forestry	1,561	None specified	None
Fisheries	None specified	None specified	None
Water accounting data (MCM)[b]			
Gross inflow	5,993	940 (surface); 160 (groundwater)	38,678[c]
Net inflow	6,120[d]	1,100	39,319
Available water – present	471	1,100	37,319
Available water – potential	5,110	1,100	36,419
Depleted fraction	2,007	600	1.02 (DF available)
Outflow	3,575	500 (only in winter)	1,100
Process fraction	249	500	0.43 (PF available)
Non-process	1,560	75	22,041
Beneficial			
Non-beneficial	197	25	0
Uncommitted	3,102	0	–900
Committed	473	0	2,000

[a]Includes 791 MCM used within the basin and 560 MCM transferred to cities outside the basin.
[b]Water accounting data obtained following procedures set out in Molden (1997).
[c]Figures for the Lerma–Chapala basin are for 1999.
[d]Includes tailwater inflow from HEP scheme on adjacent river system.
MCM, million cubic metres.

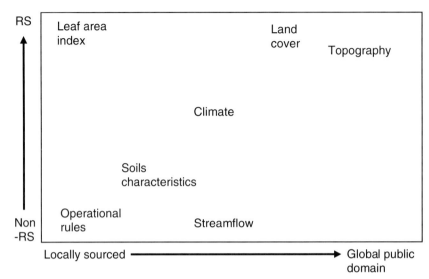

Fig. 4.4. Data required for integrated basin hydrological modelling as a function of availability and method of observation (Kite and Droogers, 2000).

Table 4.4. Summary of data sources for Gediz river basin model.

Data type	Data source
Topographic analysis	Downloaded from the Internet a Digital Elevation Model (DEM) from the United States Geological Survey (USGS) dataset; calculated streamflow travel distances using the TOPAZ (Topographic PArameteriZation) programme
Land cover classification	Base data of Normalized Difference Vegetation Index (NDVI) images taken from the NOAA-AVHRR satellite sensor at 1-km resolution
Leaf area index	Derived from NDVI data for land cover classification
Soils	Used a locally available soil map (1 : 200,000) scale as the FAO world soil map is too coarse at 1:500,000 scale
Meteorological	Used local climate station and global datasets.
Streamflow	Used local river gauging station data
Regulation	Used records of actual system regulation rules

In contrast to the above, information for analysing the institutional arrangements governing the river basin can only be collected 'on the ground'. Timelines and responsibility charts are useful analytical tools in this context. Timelines chart the development of water resource-related events within the river basin. These events are both physical, such as floods, droughts, disease outbreaks, etc., and institutional, such as enactment of water laws, establishment of water users associations, public outcries, etc. The case studies presented in Chapters 6–11 offer examples of timelines showing how physical changes within the river basin over time have been met with institutional changes to help manage and alleviate the problems encountered.

Functional responsibility charts, as described in Chapter 1, can be used to identify key actors in the water resource sector and to assess their role in relation to the nine key basin management functions. These charts can be used to structure an assessment of the data and information processes within the river basin by asking the following questions related to each function:

- What data does each organization collect?
- What processing and analysis is done on the data collected?

- How are the data stored?
- What reporting and dissemination is carried out?
- Who uses the data?
- What action and decisions are made using the data?

As with the institutional processes outlined in Chapter 1, this analysis helps to identify gaps, both in the data collected, processed and analysed, and in its use and dissemination. As will be seen in the case studies, there are significant differences in the efficacy of the data and information processes in the six river basins studied.

4.6 Irrigation System Level Information Needs

Data are required by the managers of the irrigation system to facilitate management, operation and maintenance of the system. Table 4.5 identifies some illustrative data requirements and their use for LCIS in their early growth stages.

Data may also be required by external organizations, typically a government agency, which has the responsibility of monitoring the performance of the LCIS. This monitoring can be for three reasons:

1. To collect data for regional and/or national level statistics;
2. To ensure that the association is being correctly managed financially;
3. To ensure that the irrigation and drainage system is being adequately maintained.

In the initial stages of irrigation management transfer (IMT), LCISs are often required to continue previous government agency practices of reporting cropped areas, yields and other data for preparation of regional and/or national statistics. In addition, the government has a duty of care to the farmers to ensure that their systems are properly managed, and that the fees they pay are correctly used. For this purpose, the government will usually carry out annual audits of the organization's accounts. An additional exercise that is not yet well established is the periodic auditing of the maintenance of the system's assets – the canals, drains and structures. In many cases, even following IMT, the infrastructure still belongs to government. Procedures are therefore required to ensure that this infrastructure is not allowed to deteriorate to a level where the system becomes unusable.

The extent and nature of the data collected will vary from scheme to scheme, depending on the general level of literacy, technical development of the irrigation systems, and technological capacity of the farming community. Farmer-managed schemes in Nepal, for example, generally have little formal data collection, with an equitable supply-orientated water allocation system being controlled through the use of proportional dividers. In locally controlled, professionally managed systems in Turkey, on the other hand, where water is allocated on demand and flow adjustments made using undershot sluice gates, relatively sophisticated data-collection and processing systems are required. In the Nepal case, the irrigation service fee is associated with the irrigable area, whereas in Turkey the fee is associated with the area actually irrigated and the crop type.

As pressure on the water resource increases, measurement of performance becomes increasingly important. A popular version of performance assessment currently is *benchmarking*, whereby the performance of different schemes is compared, and internal performance over time monitored. Table 4.6 presents performance indicators proposed by Malano and Burton (2001) to facilitate performance benchmarking, enabling managers to compare the performance of their system with that of other systems, in particular those that are performing well.[1] By tracking these

---

[1] Systems that are identified as performing well in relation to key indicators can be studied and the practices and procedures contributing to this performance identified. These 'best practice' activities can then be replicated by other systems in order to raise their level of performance.

Table 4.5.   Typical data needs for locally managed irrigation systems.

Data	Purpose/use
Management	
Number of farmers in command area	Indicates the maximum potential number of members in the LCIS organization. Not all farmers will necessarily be members
Names and addresses of association members	Basic data for organization membership
Revenue collected, by source	Includes revenue from each member and non-members, plus other sources of income (rents, hire out of equipment, etc.)
Management costs	Costs associated with running the association – salaries of office or administrative staff, office rental, equipment purchase or hire, stationery, etc.
Command area serviced	Potential area that can be irrigated. Provides base for analysis of performance (i.e. yield, kg/ha; fee recovery rates, $/ha; water delivery, m³/ha)
Operation	
Total seasonal or annual volume of water received	Total seasonal or annual volume of water received by the organization, based on daily measurements at the intake(s) to the irrigation system
Total seasonal or annual volume of water delivered	Total seasonal or annual volume delivered by the organization based on measurement of discharge at delivery points or estimates of volume delivered. Should correspond with irrigation service fees recovered if fees are levied based on volume delivered. Relatively few LCIS have measurement at delivery points
Operating costs	Costs associated with operation – fees for water (from bulk seller), operation equipment (motorbikes, bicycles), salaries for water masters, electricity costs for pumping, etc.
Maintenance	
Inventory of assets and their condition	Inventory of all irrigation and drainage infrastructure and its condition. The inventory can be presented in the form of tables and schematic diagrams
Maintenance requirements	Periodic or annual inspection of the irrigation and drainage system to identify the maintenance needs, costs and priorities
Maintenance costs	Costs associated with maintenance – maintenance staff wages, maintenance contract costs, equipment hire or purchase, etc.
Production	
Seasonal and annual area irrigated, by crop	Basic indicator of performance. If the area irrigated is low there may be potential for increasing the area irrigated and thus the fee recovery. Shows performance over time
Total value of crop production	Useful indicator to assess the cost of irrigation relative to the production benefits

LCIS, locally controlled irrigation system.

indicators (such as total value of fees collected) over time, it is also possible to assess the performance of a single system over time. Figures 4.5 and 4.6 present data collected from 46 irrigation schemes in Australia, showing the range of values of water delivered and costs recovered. Both types of comparisons give system managers the information (and incentives) they need to raise their performance.

4.7 Issues

Some of the most important issues related to the collection, processing, analysis and

Table 4.6. WUA and Federation performance benchmarking indicators (from Malano and Burton, 2001).

Indicator	Definition	Remarks
Financial		
Cost recovery ratio	$\dfrac{\text{Gross revenue collected}}{\text{Total MOM cost}}$	*Gross revenue collected:* Total revenues collected from payment of services by water users *Total MOM cost:* Total management, operation and maintenance cost of providing the irrigation and drainage service excluding capital expenditure and depreciation/renewals
Total MOM cost per unit area ($/ha)	$\dfrac{\text{Total MOM cost}}{\text{Total command area serviced by the system}}$	*Total MOM cost:* Total management, operation and maintenance cost of providing the irrigation and drainage service excluding capital expenditure and depreciation/renewals *Total command area serviced by the system:* The command area is the nominal or design area provided with irrigation infrastructure that can be irrigated
Revenue collection performance	$\dfrac{\text{Gross revenue collected}}{\text{Gross revenue invoiced}}$	*Gross revenue collected:* Total revenues collected from payment of services by water users *Gross revenue invoiced:* Total revenue due for collection from water users for provision of irrigation and drainage services
Production		
Irrigated crop area ratio	$\dfrac{\text{Total annual recorded irrigated crop area}}{\text{Total comm and area serviced by the system}}$	*Total annual recorded irrigated crop area:* The total irrigated area for which irrigation fees have been paid or invoiced during the year *Total command area serviced by the system:* The area provided with irrigation infrastructure that can be irrigated

continued

Table 4.6. *Continued.*

Indicator	Definition	Remarks
Water management		
Total annual volume of irrigation water delivery (m³/year)	Total volume of water delivered to water users over the year or season	Measured at the interface between the irrigation service provider and water users
Annual irrigation water delivery per unit command area (m³/ha)	$$\frac{\text{Total annual volume of irrigation water inflow}}{\text{Total command area serviced by the system}}$$	*Total annual volume of irrigation water inflow:* Total annual volume of water diverted or pumped for irrigation (not including diversion of internal drainage) *Total command area serviced by the system:* The area provided with irrigation infrastructure that can be irrigated
Annual irrigation water delivery per unit irrigated area (m³/ha)	$$\frac{\text{Total annual volume of irrigation water inflow}}{\text{Total annual recorded irrigated crop area}}$$	*Total annual volume of irrigation water inflow:* Total annual volume of water diverted or pumped for irrigation (not including diversion of internal drainage) *Total annual recorded irrigated crop area:* The total irrigated area for which irrigation fees have been paid or invoiced during the year
Main system water delivery efficiency	$$\frac{\text{Total annual volume of irrigation water delivery}}{\text{Total annual volume of irrigation water inflow}}$$	*Total annual volume of irrigation water delivery:* Total volume of water delivered to water users over the year or season *Total annual volume of irrigation water inflow:* Total annual volume of water diverted or pumped for irrigation (not including diversion of internal drainage)
Maintenance		
Maintenance cost to revenue ratio	$$\frac{\text{Maintenance cost}}{\text{Gross revenue collected}}$$	*Maintenance cost:* Total annual expenditure on system maintenance *Gross revenue collected:* Total revenues collected from payment of services by water users
Maintenance cost per unit command area ($/ha)	$$\frac{\text{Maintenance cost}}{\text{Total command area serviced by the system}}$$	*Maintenance cost:* Total annual expenditure on system maintenance *Total command area serviced by the system:* The command area is the nominal or design area provided with irrigation infrastructure that can be irrigated
Administration		
WUA membership ratio	$$\frac{\text{Total number of WUA members}}{\text{Total number of farmers in command area}}$$	*Total number of WUA members:* Total number of farmers registered as members with the WUA *Total number of farmers in command area:* Total number of farmers with registered landholdings within the WUA command area

WUA, water users associations.

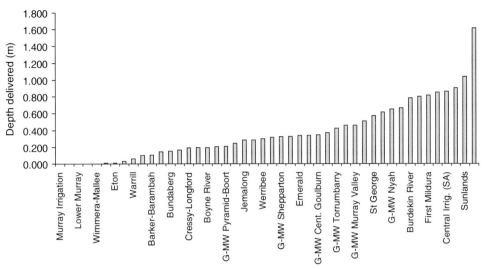

Fig. 4.5. Example of comparative plot of irrigation water delivery for different irrigation systems (ANCID, 2000).

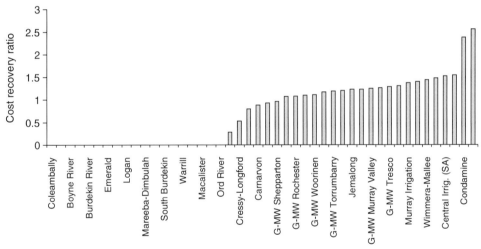

Fig. 4.6. Example of comparative plot of cost recovery ratios (ANCID, 2000).

use of data and information in a river basin context are summarized below.

4.7.1 Issues at river basin level

1. *Coordination and sharing of data.* As basins develop, different agencies become involved to different degrees in water resource data collection, processing and analysis. There is often a lack of coordination, cooperation and sharing of data between these agencies. As the pressure on the available water resources increases, improved cooperation and data sharing is needed between these agencies, and between collecting agencies and interested third parties such as farmer organizations. Measures to enhance co-ordination and sharing can include the formation of river basin councils, and the creation of national data banks for water resources.

2. *Adaptation to changing needs.* Government organizations can be slow to adapt to a changing environment and to changing needs. Change is often prompted by crisis events, but managers should be alert to the need for change before a crisis occurs. Rising levels of pollution in rivers and falling groundwater table levels in river basins are two areas where, typically, action is not taken in good time to alleviate looming crisis. In these circumstances, it is the responsibility of managers to gather, present and disseminate data to highlight the issue and stimulate a response.

3. *Computers, remote sensing and GISs.* There is increasing use of computers for data storage, processing, analysis and presentation. They have a valuable role to play in presenting data in ways that make them meaningful to non-professionals, and a central role in processing and analysing complex data, as in water resources modelling. GISs are now widely used to store and present data, and such uses can be expected to expand. Increasingly computers are being used in association with remotely sensed images to analyse and present complex relationships, allowing analysis in forms hitherto not possible.

4. *Water rights.* When water supplies are abundant, it may not be necessary to specify water rights or licences for specific uses. As water becomes increasingly scarce, society has to decide who should have access to the available water and in what quantity and quality. Users must be assigned water rights so that they can plan for the future, particularly in relation to making investments. Information on water resource use and availability over time is vital in allocating water rights and in determining procedures for real-time allocation of water in times of shortage. Water right allocation and reallocation can be politically contentious, but become increasingly necessary as basins mature and close.

5. *Pollution.* Significant pollution is often a feature of river basins in the middle stages of their development, as water resources are taken up and begin to become scarce. Pollution is not only hazardous to human health and the natural ecology; it requires large amounts of water to dilute it. Many river systems have had to die before adequate action has been taken to remedy lethal levels of pollution. The River Thames flowing through London was at one time so polluted that the Members in the adjacent Houses of Parliament had to recess during the summer months to avoid the stench. Control of point and non-point sources of pollution requires legislation, funding, and political and social will to achieve.

6. *Over-abstraction of groundwater.* A feature of the latter stages of river basin development, over-abstraction of groundwater, appears to be one of the most difficult issues for societies to address. In the case of the Lerma–Chapala (Chapter 8) and California's Central Valley (Chapter 7), both well-developed river basins, over-exploitation of groundwater remains a major unresolved issue. It is a classic 'tragedy of the commons' scenario, on which society has to form a consensus to legislate and abide by.

7. *Enforcement.* Data and information are the basis for establishing and enforcing regulations and standards related to water resources management. In some societies, despite adequate legislation and data, there is a lack of political or social will to enforce the legislation. In such contexts, information and its wide dissemination are key factors in raising awareness of the problems and building the political and social constituencies to support legal enforcement.

4.7.2 Issues at LCIS level

1. *How much data to collect.* The type and quantity of data collected will vary widely among countries and with size and type of irrigation system. There is often a noticeable difference between the kind and amount of data that water resources and irrigation professionals think is required for management purposes and that which farmers see as being necessary. Systems large enough to employ professional managers, as is typically the case in Spain, Mexico and Turkey, for example, will generally have more comprehensive management information systems than schemes that are managed by farmers without professional training in water resources or irrigation management. Generally, larger schemes, especially those growing principally for a market, require more comprehensive quantitative information systems than smaller, simpler schemes.

2.   *Discharge measurement.* Whether discharges are measured and the data used or not marks a major divide among irrigation schemes. For smaller simpler schemes, there will not be significant benefit to the additional work involved in measuring and using discharge data, while in schemes where water is scarce and returns to irrigation are high, significant benefits can arise through improved management based on discharge measurement. On large irrigation schemes, measurement becomes essential if water is to be distributed equitably. A reliable discharge measuring system is not easy to establish and sustain, requiring measuring equipment or structures, calibration, maintenance, regular data retrieval and processing, and trained staff.

3.   *Data processing and analysis.* There is a marked difference between data processing and analysis with and without a computer. Where LCIS managers do not have a computer, data processing will necessarily be more basic. Nevertheless, useful information can be generated from simple processing of the available data using a hand calculator. Once, again, the size of the scheme, the value of its output, and the presence of professional managers will determine whether computer-based data processing is appropriate.

4.   *Transparency.* Access to information about system operations is essential for good governance and management. When farmers' access to information about water availability, rules, deliveries and finances is obstructed, corruption and misuse of funds and resources becomes easier and more likely. To ensure transparency, clear requirements for information sharing need to be included in water legislation governing LCISs. All farmers should be aware of their right to inspect the association's account books, and information on the scheme's physical and financial performance should be presented and discussed at the association's annual general meeting. Government regulatory authorities should ensure that: (i) farmers are aware of their rights; (ii) annual financial audits are conducted; and (iii) the LCIS managers organize annual or semi-annual general meetings and present the requisite performance information to farmers.

5.   *Access to external data.* LCIS have an important role to play in the management of the larger basin, defending irrigators' interests and reaching agreements with other water users over water sharing and water quality issues. To play this role effectively, LCIS managers need access to basin-level data collected by government agencies and others, and the ability to use and interpret that data. This requires guarantees of access to information, and the professional skills to understand, discuss and debate that information. Here a federation of LCIS in a particular basin can be extremely valuable in pooling resources needed for effective representation.

6.   *Training for LCIS personnel.* Training is a crucial and underemphasized part of the turnover process. In project-funded transfer programmes, training is often a part of the assistance package provided. Seldom, however, do project sponsors or governments establish an on-going training capability for LCIS following IMT. As elected leaders and managers rotate following elections, skills from a one-off training effort are lost and management quality degrades. Here too, a federation of LCIS can be important in pooling resources to support a national training programme for member organizations.

4.8 Conclusions

Data and information are fundamental to the good governance and management of a river basin. In keeping with changes in society in general, decision making for river basins is increasingly being devolved downwards from national to state or regional governments and local stakeholders. Associated with this trend is an increasing awareness amongst the general public of water-related issues, and their rights and responsibilities in relation to the use and management of increasingly scarce water resources. Experience in many countries is showing how the sharing of data and information by government agencies with the wider public enables management to move on to higher performance levels through enlisting support and collaboration from water users. Societies and government organizations that seek to hold on to power through restricting the sharing and access to water resources data and information will find that they are limiting their capability to make the most productive use of the available water resources.

In water-scarce situations, accountability and transparency through provision of data and information at both the river basin and irrigation scheme level enable management to gain the consensus and support required from water users to address issues related to scarcity. Formulation, agreement, compliance and enforcement of measures to manage within the context of scarcity are empowered by the acquisition and sharing of data and information between stakeholders. For LCISs, the acquisition and sharing of data become central to their need to manage their external environment to secure their water rights and gain a fair share of the available resource. Within their irrigation schemes, improved data and information processes are required as part of the pressure placed on the irrigation sector to improve management and justify its share of the water resource.

Data and information needs change over time, and those responsible for the collection, processing and analysis of data need to review their information strategies and information systems to reflect the dynamic environment in which they operate. As river basins develop, data and information needs both broaden (to encompass more diverse requirements) and deepen (to enable more detailed resource assessment, allocation and monitoring). The availability of microcomputers and associated software are dramatically changing the possibilities for river basin managers, allowing them to access and process data to a level hitherto only possible in research establishments or national headquarters offices. Spatial representation systems, such as remote sensing and GISs, are becoming standard tools for system analysis, allowing more precision in the analytical processes associated with river basin management.

References

ADB (1996) Towards effective water policy in the Asian and Pacific Region. Arriëns, W., Bird, J., Berkoff, J. and Mosley, P. (eds) *Proceedings of the Regional Consultation Workshop*, Manila, 10–14 May, Asian Development Bank.

ANCID (2000) *1998/99 Australian Irrigation Water Provider – Benchmarking Report.* Australian National Committee on Irrigation and Drainage, Victoria, Australia.

Kite, G. and Droogers, P. (2000) *Integrated Basin Modelling.* IWMI Research Report 43, International Water Management Institute, Colombo, Sri Lanka.

Malano, H. and Burton, M. (2001) Guidelines for benchmarking performance in the irrigation and drainage sector. International Programme for Technology and Research in Irrigation and Drainage (IPTRID), Rome.

Molden, D. (1997) Accounting for water use and productivity. SWIM Paper No. 1, International Irrigation Management Institute, Colombo, Sri Lanka.

5   Financing River Basin Organizations

Charles L. Abernethy

5.1 Introduction: Uses and Demands for Water

It has become quite widely accepted that countries should be aiming towards comprehensive management of water resources through organizations based on river basins or aquifers. This chapter addresses the situation where a country has already decided that it wants to assign an organization for this management purpose. In most cases, this will mean establishing a new organization, rather than extending the functions of an existing organization. Such organizations will not develop effectively unless they can be provided with adequate financial resources. The questions considered here relate to the ways of supplying these financial resources, and the impacts that these financing processes may have on the behaviour and effectiveness of the organization, and on the people who are supposed to be helped by its existence.

In this respect, the experiences of the richer countries may not offer much useful guidance to the developing countries, especially the poorest ones. Their patterns of water use are radically different. Table 5.1 shows the breakdown at the broadest level, among the three biggest user sectors. The data for this table are drawn mainly from the World Bank's 'World Development Report' for 1998–1999. The data on the broad sectoral distribution of water uses (among agricultural, domestic, and industrial uses) are not up to date, since most of the information about this in the global reference literature dates from 1987; however, these numbers are adequate to illustrate some general patterns.

The boundaries defining the four wealth classes in Table 5.1 have been set at the following levels (1998 data): $2375, $5750 and $12,500. These are in terms of gross national product per person per year, expressed in equivalent US dollars, at PPP (purchasing power parity).

Table 5.1. Sectoral use of water. Units: % of annual freshwater abstractions.

Sector	Agriculture	Domestic	Industry
Wealth category			
Low income	89	4	7
Lower middle income	74	8	18
Upper middle income	73	12	15
High income	40	15	45

Source: Adapted from World Bank (1999).

In the rich countries, industrial users predominate. In the poorest countries, the industrial category is not yet very significant, whereas as much as 89% of all the water abstracted is used for agriculture.

These patterns of use illustrate two obvious factors that have great influence on the financing situation. First, in a basin where the majority of users are small agriculturists, they are usually extremely numerous, forming a large majority of the people in the poorest countries. Secondly, the productivity of water used in agriculture is usually very much lower than that of water used in industry. So when we deal with industry we are usually dealing with a relatively small number of people who are engaged in relatively profitable activities, while when we deal with agricultural users (in developing countries), we are probably dealing with large numbers of people whose financial resources are very meagre.

We can also note (although it is far beyond the scope of this chapter) that the low productivity of irrigated agriculture has well-known links with the agricultural production and market-access policies of rich countries, and other issues of global scale which no developing country can modify much by its own choice of policies.

Domestic water supply is different again. We cannot compare it with other types of use on the basis of productivity. Domestic use is essential for human health and indeed survival. So we supply water for social objectives as well as for productive objectives, and these are not really comparable in a financial sense.

That distinction between social and productive objectives is not as clear as we might like. The four basic human needs for water – drinking, washing, cooking and sanitation – are certainly essential, but domestic uses of water can include many non-essential uses. When we compare the consumption patterns of the rich and poor countries, this becomes very evident. As Table 5.2 shows, in the rich countries the amount of domestic water used per person is very much higher than in the poorest countries. Also, within each country there are similar variations of consumption, related to poverty or affluence.

In Table 5.2, we see that in the poorest countries the abstractions for personal use are minimal. We may estimate the basic needs at about 30 m³/person/year[1], so the figure of 16.4 indicates that many people are obliged to satisfy those needs in ways that do not reach the formal statistical system. These are people who have to bathe in open bodies of water, carry household water from local streams, and in other such ways are omitted from the data.

Table 5.2. Abstractions per person. Units: m³/person/year abstracted from the natural systems.

Sector	Agriculture	Domestic	Industry
Wealth category			
Low income	332	16.4	26.7
Lower middle income	339	36.3	81.7
Upper middle income	332	55.9	68.9
High income	386	146.5	442.2

Source: Adapted from World Bank (1999).

---

[1] This is of course an approximation, dependent on many circumstances. Various authors quote numbers in the general range 20–35 m³/person/year. See for example Seckler *et al.* (1996: 236), citing Gleick (1996) for a figure of 20 m³/person/year, or Clarke (1993: 20), estimating 35 m³/person/year. The World Bank (1992: 99) notes, in regard to health and the fourth component of basic needs, sanitation, that 'the use of water for personal hygiene usually increases only when availability rises to about 50 litres/person/day' (18 m³/person/year), which seems to imply that something less than this is the amount needed for the first three basic needs. The most basic need of all, drinking, has an extremely small impact on demand, accounting for some 0.4–0.7 m³/person/year depending on location.

Table 5.2 also shows that, around the world, the gross amounts of water extracted for agriculture are quite similar. Here again we see that, although irrigated agriculture is often blamed for water scarcity in developing countries, the amounts used for agriculture in poor countries are lower than in rich ones. The differences are small, and probably not meaningful in view of the general doubts about data validity; but, as far as the data go, they do not indicate that agricultural users are more wasteful of water in poor countries than in rich ones.

We can also relate these figures to basic human needs. At the minimal nutritional levels required for sustaining human health, in a society where the basic food is a cereal crop such as rice or wheat or maize, it takes a quantity of water in the order of 300 m^3/ person/year to grow that food, if the water is applied very efficiently.

Of course that figure does not reflect directly the abstractions from the natural river systems, since much of the crop water requirement is supplied by rain, and also most of the irrigation water is not applied at highest efficiency; nevertheless, the need to satisfy a certain basic food requirement applies to us all, and it is useful to note that this need is in the order of ten times bigger than our basic need for domestic water, and in the order of 500 times bigger than our need for water to drink.

These widely differing patterns of water abstraction and use have various implications for the effectiveness of financial policies as instruments for influencing water uses and demands. In the affluent countries, where personal domestic consumption is very much more than the amount required for satisfying the four basic needs, a charging policy may have significant impacts on consumption. A high charge may make people reduce frivolous or non-essential uses of water, or may just make them more conscious of water costs, and therefore induce them to change their behaviour, in ways such as becoming quicker to attend to leakages.

But in countries, or families, where domestic consumption consists simply of satisfying basic needs, a charging policy is not so likely to have those impacts. Supplying a basic need such as drinking is not something about which the 'user' can exercise much choice. If the cost increases, the basic consumption will have to stay roughly the same. Any financial effect on that user will appear in some other direction, by not spending on some other item that must appear as less vital.

The nearer a country, or a village community, is to poverty or to an elementary subsistence level, the more this argument will apply to the case of agriculture. People will not feel that they have the option to reduce their consumption, since food is a basic need to just the same degree as drinking water. So, while the potential effect of water charges in constraining consumption may be quite large in rich countries, it may be very small in poor countries.

5.2 Financial Productivity of Water

When we look at the productive applications of water, we can find many illustrations of the relatively low financial productivity of agricultural water. For example, Schiffler and others (1994) analysed the economics of water uses in Jordan, a country with one of the world's lowest levels of water resources per person. They reported that the average productivity of water in industry was 11.2 dinars/m^3 (about US$16.8/$m^3$ at the bank exchange rate then) whereas the average productivity of water in agriculture was 0.28 dinars/m^3, or 2.5% of the industrial level.

Within the agricultural sector, there are further huge variations of productivity. The productivity of basic cereal crops in the developing countries is usually around the equivalent of a few US cents per cubic metre, while fruit and vegetable crops, especially those for export, may show productivity as much as 100 times greater than the cereals. In the Jordan case, Schiffler et al. (1994) found that the productivity of grapes was 130 times more than that of wheat.

So here again we have the problem of basic needs. In the poorer countries, or poorer environments within any country, these low-productivity cereal crops – rice, wheat and maize especially – dominate the agricultural scene, and are not necessarily grown for the market. In studies of five small irrigation systems reported by PMI-Burkina Faso (1997), it was found that the proportion of products marketed was 25.6%. In the system with least road access, this fell as low as 5.3%. The rest was household consumption.

To the users in those villages, the financial productivity of water must seem an utterly irrelevant concept. The idea makes sense (to the user of water) only in the context of marketable alternatives. Of course there is a governmental or 'public interest' viewpoint that may suggest something else, concerning optimal uses of a scarce national resource. But if we feel that we may be moving towards any kind of user-based financing system, it seems that we have to try to understand how these things seem from the users' perspective.

Briscoe (1997: 341–342) has emphasized particularly the idea that there can be a high opportunity cost associated with agricultural uses of water. The nub of this argument is that, where water is not abundant, low-productivity applications of it, for example to grow cereal crops, deny that water to higher-value potential users. He points out that prices charged to agricultural users are, typically, around 10% of those charged to urban and industrial users for comparable volumes delivered.

This is an argument that is easier to act upon in a mixed agricultural/industrial economy (or 'middle-income' economy), such as some of those in South America; but it seems to carry less weight in countries where 89% of the utilization is agricultural and the opportunities for transfers of use are correspondingly few. The concept of opportunity cost depends on the existence of such opportunities.

On the other hand, it is not safe for poor countries to treat the opportunity cost argument as irrelevant to their situation. In recent times, several countries in east and South-east Asia (Thailand is an example) experienced rapid economic growth, continuous over more than a decade. A

consequence of this was the arrival of many new investment opportunities, some of which would depend on transferring water from a traditional low-productivity use into one of the new uses where its value (in economic productivity terms) would be some orders of magnitude more. In the absence of sound institutional mechanisms for responding to these opportunities through orderly, voluntary transfers and compensation of the prior users, these changes have occurred, but sometimes at high social cost.

Perhaps the productivity dimension can be summed up somewhat as follows (referring in particular to countries in the lower parts of the national wealth tables): people are not ready to treat the major uses of water – domestic, agricultural, industrial – on a basis of simple financial equivalence. There is a quite widespread view that agricultural and (especially) domestic uses should not be compared against industrial uses in revenue-per-cubic-metre terms. Even if there is acceptance that agricultural water is often distributed and applied inefficiently, the idea that water used agriculturally should be transferable to industrial use, just through market pressures, does not seem to be generally acceptable at this time. Political resistance to introduction of agricultural water charges, in democratic developing countries such as Sri Lanka and Thailand has provided evidence of this in recent years. On the other hand, as agriculture moves from subsistence to commercial modes of production, this attitude will no doubt undergo some change.

5.3 Components of Costs

The financing question, for river-basin organizations, depends of course on the tasks that each country may decide its river-basin organizations should perform. The scope of basin organizations falls into three broad categories, which may overlap in some countries:

● *Regulatory organizations:* organizations which supervise, regulate access and monitor the management of water, and make rules which service-providing organizations have to follow, but have no direct role in service provision;

- *Infrastructure managers:* organizations which own the principal structures and facilities for water supply, and invest in new ones, but do not provide water supply services directly;
- *Service providers:* organizations which deliver water supply services directly to users or customers.

It may appear that the first of these, regulation only, is a relatively cheap alternative. If we adopt that kind of organization, it may seem that the budgetary issue will be small and easily manageable. But that is not the case. If regulation is to be done well, it needs a significant amount of finance. A short list of the primary regulatory functions would include:

- monitoring of the quantity and quality of water in all rivers and other natural water bodies in the basin;
- conserving and protecting the watershed;
- making rules about abstraction, uses, disposal and pollution;
- supervising the application of a system of water abstraction rights or licences;
- ensuring compliance with rules through monitoring of activities, public information programmes, court processes, etc.; and
- conducting transparency and accountability programmes, to ensure wide and continuous understanding, acceptance and support of basin management principles.

These tasks amount to a quite formidable financial commitment. They are most urgent in basins where water resources per person are already low. On the other hand, an organization that has no service delivery function does not have a direct customer base from which a proportion of funds can be sought. These considerations show that the design of basin organizations cannot be separated from the question of how they will be financed, at a level sufficient for

them to discharge the tasks that are assigned to them.

The movement towards establishing river-basin management organizations is coming at a time when governments, in both rich and poor countries, have been trying to reduce the amounts they budget for providing water services; so the idea that these organizations might be funded from the budget of a central government ministry may not be received well in many countries. In the developing countries, the main feature of this trend has been the numerous programmes of irrigation management transfer, which began in a few countries such as the Philippines and Colombia in the middle 1970s and have since become very general, indicating a widespread perception that central subsidizing of water services is difficult to sustain.

The experience of irrigation management transfers in the past 20 years has however shown some of the difficulties that can occur when governments try to transfer the responsibility for certain tasks and their related expenditures from a service-providing organization (such as a government irrigation department) to an organization of service-receivers.

We can distinguish four kinds of costs that are faced by water organizations which are service providers:

- capital investment (constructing facilities for capture, conveyance and distribution of water; purchasing equipment; providing the buildings and other hardware of the management systems);
- major repairs and renewals of equipment and infrastructure;
- direct recurring costs (operation and maintenance);
- overhead costs (sustaining an administrative structure, including probably higher and remoter organizational levels, national, regional, etc.).

An economist might say that the first two of these belong together as one 'capital' category. But it seems better to make them distinct, as they usually happen far apart in time, and by the time that the need for renewals becomes urgent, the fact of the initial capital investment has usually caused great changes in the economic condition of the users.

In irrigation management transfer, governments typically aim to transfer to organizations of the service receivers the responsibility for some or all of the third cost category, operation and maintenance, but usually the first and fourth categories are not transferred. The responsibility for the second category is often left unclear, and has been a source of problems in a number of such transfer programmes, because it creates doubt about the borderlines between the two parties.

The overhead costs of governmental irrigation organizations are not often discussed in the relevant literature. This could be because they are very large. Especially in Asia, government irrigation organizations are among the strongest and most long-lived organizations, and have developed large superstructures, often based in capital cities far away from their client populations. This seems to make the overhead cost a special one, which is not likely to be transferable to the individual end-users.

5.4 Regulation and Service-delivery Functions

Let us look now at the three different modes of basin organization which were identified at the beginning of the preceding section. In developing countries, we can usually find existing organizations that exercise the functions of service delivery for each specific use category. These are often quite old organizations, which have developed a variety of specialist skills and have large professional work forces. It seems unlikely that governments will abolish them. It seems unlikely, therefore, that basin management organizations will evolve towards direct service provision to the ordinary citizens. A more probable path of evolution will be towards basin organizations taking up the regulatory functions, while direct service delivery will remain the task of other organizations which manage urban

water supply, agricultural water supply, hydropower and other specific services to people, to companies and to other user organizations.

In theory, then, the service-providing organizations should become more clearly service-oriented, should behave more commercially, should become more subject to compliance with laws about pollution and other adverse social consequences of their activities, and (depending on the politics of the country) may be considered for privatization; while the regulatory organization exists in the public domain to ensure good laws, allocation of resources by administering a water-rights or licence system, conservation and protection of water sources, and compliance with all of this.

That still leaves open the very difficult question of who should undertake new capital investments for infrastructure provision. Will it be the service providers, or the basin organizations? There are strong arguments both ways. But it seems clear that this issue will be a vital one in determining the character of a basin organization, and its relationships with service-providing organizations. If basin organizations are going to be constructors of major new facilities, their financial requirements will be much heavier than if they are purely regulators.

It seems that the primary reason why we need basin organizations, as the prospects of water deficits appear in an increasing number of countries, is for establishing compliance with a body of rules that will enable the people at large, through institutions, to regain some kind of control over the diminishing quality and quantity of water in their rivers. If we take that view, then perhaps we will think that this is a sufficiently huge and important task, and that we should not give the same organization more conventional tasks, such as construction of major facilities, or even ownership of facilities that exist already.

One of the problems of establishing basin-management organizations is that there is clearly a potential conflict with existing organs of local government, which almost everywhere have boundaries that are different from the boundaries of river basins.

It is said sometimes that, since provinces or other local government units are responsible for achieving development within their specific boundaries, they must have control over such a major development factor as water. There is certainly much force in that argument. However, the separation of regulatory and service-delivery functions opens a way to escape from this problem. It is possible to organize regulatory functions on a river basin (or aquifer) basis, while service-delivery functions can be organized on a different basis which may conform more closely to the boundaries of local or provincial governments.

5.5 Sources of Funds

The Dublin and Rio de Janeiro Conferences of 1992 enjoined us to regard water as an economic good (International Conference on Water and Environment, 1992). That seems to mean that users of it should pay for it according to the amount that they use. The way ahead, according to this view, seems to involve finance coming primarily from users of water, paying to service-providing organizations. In that pattern, it would seem practicable to finance regulatory basin organizations through some system of levies on the income of the service-providers.

On the other hand, it is difficult to see that service-providing organizations, in the poorer countries, are going to be able to behave commercially, and at the same time invest substantially in new capital facilities. The low profitability of the prime user, irrigated agriculture, together with the socio-political resistances to full-cost charges for basic needs (discussed above in Section 5.2) indicate that, for many countries, this is not an immediate prospect.

Probably, too, the phrase 'economic good' suggests that the prices we pay for water services should somehow reflect the sort of factors that usually influence prices of other economic goods. For example, if water is an economic good we might expect its price to rise in times and places of scarcity, and to fall in circumstances of poor quality.

Concepts like this, however satisfactory they are economically, face many difficulties from social and political angles. Water has been treated for long as an aspect of welfare provision, and in many places long periods of provision of agricultural water at zero price, or extremely low price, have promoted high effective demand, which is now very difficult to reduce.

However, there seem to be few alternative routes available for financing the activities. Either they must be financed from user charges, or they must be financed from central government budgets, or they will probably not happen effectively at all. The problem of central government funding, for the poorer countries at least, is that there is very little of it available, and there is strong competition for that little amount. We can see from the fate of (for example) hydrological data-collecting organizations, which in many countries have become weak and inadequate for their tasks, that centrally funded organizations which are doing things that do not have direct popular appeal are likely to be left on the sidelines in the budget contest. Funding river-basin organizations this way may well make them unstable, and unable to pursue consistent long-range planning.

In Europe, there has been a trend in recent years towards the use of abstraction licences as a means of raising a significant proportion of the funds needed for sustaining regulatory organizations. This becomes possible when the regulatory function is clearly separated from the service-provision function. Regulatory organizations assess the available quantities of water and issue licences accordingly. In that system, the service-providing organization can be just another holder of an abstraction licence.

Abstraction fees are not the same as user fees. A service-providing organization may pay abstraction fees to the regulatory organization, and then sell the water to ordinary people or businesses, charging them a user fee, which exceeds the abstraction fee in order to cover the costs and financial risks of delivering the water. In such systems, abstraction licence fees may be graded according to scarcity.

Buckland and Zabel (1998) describe the workings of these systems, and report abstraction fees that are typically around the equivalent of 1–2 US cents/m^3, but in some cases significantly more. In some countries the product of abstraction fees is sufficient to cover the cost of all regulatory functions.

In a licence fee system, there are two ways of charging the user. The charge may be based on the measured actual consumption of water, or it may be based on the amount allowed by the licence. The system of measuring actual amounts involves a higher level of regular metering of the users, whereas the licence-amount method can be implemented with only occasional checks, to ensure that the conditions of the licence are not exceeded. Dual systems, combining a fixed quota and a volumetric charge for excess over this, are also used (see for example, Tardieu, 2001).

If the cost of abstraction licences is set high enough, they can have an effect on the consumption of water. In the German state of Hamburg, for example, a relatively high abstraction fee for ground-water licences caused about a third of the licences to be renounced, and handed back to the regulatory organization, which could then re-issue them to others (Buckland and Zabel, 1998: 270).

There are other possible sources of revenue for a basin organization. We can note three principal areas:

- waste-water disposal charges;
- pollution charges or fines;
- charges for permits for other water-based activities, such as fishing, navigation, recreation, etc.

Many of the charging methods noted above can be levied by service-providers, but would not easily be collected by a purely regulatory organization, since it has no retail delivery function and is not in regular direct contact with household and farm users of water. Therefore, if we want to establish a basin organization of the regulatory type, it seems that we should seek financing methods that are based mainly on the relationship between the regulator and the service-providing organizations.

The problem of dealing with numerous individual small users of water is a real source of difficulty in many developing countries. Licence systems, such as for example licences to install and operate ground-water wells, should therefore be introduced only after careful preparation, as they may not be respected if the capacity to ensure compliance does not exist. However, licensing systems do not involve the same amount of administrative work as volumetric use fees, and (subject to the condition of adequate enforcement capacity) can have a useful place in the scheme of regulatory finance.

The principal types of mechanism for funding a purely regulatory basin organization therefore appear to be:

- abstraction licence fees,
- waste-disposal permit fees,
- pollution charges and fines,
- permits for water-based activities,
- government subsidy.

The above discussion has focused on the funding of recurrent costs: regulatory and operational costs. The question of sources of funding for capital costs presents other issues which will not be analysed here. However it is worth noting the data presented by Briscoe (1999: 307), who showed that, in a sample of five developing countries, the average share of the private sector in water infrastructure investments was only about 6%, compared with an average private share of about 49% in investments across other infrastructure types: energy, telecommunications and transport. The low level of private investment in water reflects the low financial returns that it produces, especially in its major developing-country application, irrigated agriculture. For this reason, we should expect that capital investment in water is likely to remain predominantly a public-sector responsibility, in the medium term at least. The problems of bringing river-basin organizations into this aspect are noted further below (Section 5.8).

5.6 Methods of Assessment

Funding of regulatory organizations from abstraction charges, direct user charges, or from a percentage levy on the user charges collected by the service-providing organizations does not necessarily mean that all categories of users pay at similar rates. When we examine current charging practices, world-wide, we find a tendency to charge agricultural users much less, and industrial users much more, than the average.

This leads us to the question, how should charges be assessed? If basin organizations are to draw their funding ultimately from user charges, how will those charges be calculated?

This is related to other issues, about the impacts that we may want a charging system to have upon patterns of water consumption. It also brings in some very complicated issues related to the quantity, quality and locations of disposal flows, returning to the natural system after use.

Industrial users are accustomed to pay for measured quantities of water delivered to their premises. Urban users in the better-off suburbs also probably pay on the basis of measured volumes, and poorer users, especially the very poor, also pay, though probably not for measured flows but for volumes brought by water-carriers.

But the biggest users in poor countries, the farmers, generally do not pay by volume at all. In the countries where irrigation service fees are levied, the overwhelming majority pay an amount that is based on land area. There are many variants of this, such as seasonal differences, crop differences, and so on; but the central point about the dominant current practice is that the marginal price, the cost to the farmer for taking more water, is normally zero.

Countries vary in the way they account for water that flows back to the river systems after use. In virtually all the uses of water, there is some 'return flow', but the amount of this varies, and in many uses it is difficult to measure it. Briscoe (1997: 345) says that:

> taking the US as an example, consumptive use as a percentage of withdrawals was 56% for irrigation, compared with 17% for

urban water supplies, 16% for industry and just 3% for thermoelectric power.

The UK, following the logic of these different levels of consumption and return, adopted a classification of use types into four bands, according to their average proportions of return flow. In such a system, users who consume a large fraction of what they abstract (such as irrigation) are charged more heavily than users with a high return percentage (such as power generation). There are of course quality aspects in relation to these return flows as well, which can be dealt with by the different mechanism of pollution charges.

In developing countries, water charges do not vary, generally, according to scarcity of the commodity. Water prices are usually calculated on some basis that is related to the cost of delivering it. That means that it stays the same, and does not respond to variations in available resources. In many countries, charging scales are centrally or provincially determined for large sets of irrigation systems, so that systems with water abundance and systems with local water scarcity are obliged to charge their users the same price.

Indeed, when we look at inter-country comparisons, although the variations in charging practices are enormous, there are signs of a correlation between scarcity and price policy, but it is a correlation that is opposite to economic logic. Some of the lowest charges (even when comparisons are based on purchasing power parity) are found in dry countries such as Egypt or Iran, while high charges can be found in much wetter places. This presumably represents a socio-political logic, which may well be stronger than economics.

There are even cases where the cost of taking water for irrigation becomes lower in the driest, hottest time of the year. The middle Niger river is such a case, simply because at that time the river level is comparatively high and the cost of pumping water to adjacent land is therefore less. This kind of anomaly results from basing charges only on the cost of service delivery, which is effectively unrelated to scarcity, and often is only weakly related to quality, or to demand.

A river-basin organization could reduce some of these anomalies. In many countries, water has been made legally the property of the state. It is possible, therefore, for a river-basin organization to charge the service-providing organizations on the basis of the measured amounts that they extract from the natural system (as in the European examples of abstraction licences, noted earlier). Each basin organization can devise its own level of charge, related to the amounts of water that it has available for abstraction. It is possible for those charges to be varied along some seasonal or even monthly scale. In this way, a basin organization could exert some pressure on the service-delivery organizations to look for ways of moderating their rates of water use, while at the same time improving its own financial independence.

5.7 Collection

In the agricultural sector of developing countries, the problems of how to finance irrigation services and how to collect irrigation service fees from users of agricultural water have been prominent issues throughout the 1980s and 1990s. It cannot yet be said that the issues have been satisfactorily resolved. This experience should make us aware that the establishment of new basin organizations in developing countries is going to face similar difficulties.

Financing urban domestic water supplies is not any easier than financing irrigated agriculture. The World Bank (1993: 126–127) reported its experience in lending for water projects in these terms (referring mainly to non-agricultural uses):

> The Bank has maintained the policy that cost recovery should be sufficient to pay both for operations and maintenance and for a fair return on capital investment . . cost recovery was rated as unsatisfactory in 80 of 114 projects. And, in 78% of countries receiving water supply and sanitation loans, financial covenants were not fulfilled. In 49 of the 120 water supply and sanitation projects, fees were not raised enough to meet financial requirements due to government constraints.

We can perhaps make a guess, that these problems, of investment projects whose cost-recovery conditions are not implemented in reality, happen because such projects are prepared in the bureaucratic domain, and subsequently meet strong resistances in the political domain, due to neglect (in project preparation) of the weight of the people's views. This can only increase, as more countries are inclining towards democratic modes of government.

The compliance problem, in respect of irrigation service fees, became famous during the 1980s. The Philippines, especially, made the 'viability index' a central feature of its institutional reforms: this index is the ratio between fees collected and the costs of operation and maintenance. Field officials of the government agency could receive bonuses depending on the percentage of fees actually collected. (Oorthuizen and Kloezen, 1995: 18; Svendsen, 1992: 5). This strategy addressed financial viability, but has been criticised by various observers as having a negative effect on (especially) maintenance, since reduction of expenditure was one way of enhancing the viability index, at least temporarily.

Studies of the costs of fee collection show that they can be a significant proportion of the total amount collected. 'Passive' collection, meaning the kind of system where each user is expected to bring the fee to the collection office, seems prone to abuses, or at least to long delays in payment, which present serious cash-flow difficulties to service-providing organizations. 'Active' fee collection, using paid collectors who visit houses or farms, incurs a significant wage cost. Both methods need accounting staff and certain facilities.

It is hard to make any general statement about the proportion of fee income that has to be used for fee-collection (that is, to ensure that the organization's income flow will be stabilised) because of the inherent variability of the process. For example, Oorthuizen and Kloezen (1995: 28) found that the average cost of fee collection in a small Philippine system was 10.8% of the organizational income, but even in that one system this percentage varied between 5%

and 15% over their 6 years of observations. These percentages are clearly linked to the degrees of motivation and energy in the collecting organization; this is doubtless why the Philippines introduced collection incentives. In other cases there may be indifference within the bureaucracies, and resistance among the water users; both of these would be likely to result in collection costs being substantially higher in proportion to income. Cases have even been noted where collection cost is of the same order as the income that the collection yields.

In the Niger systems passive collection was used. Costs of collection were still significant, as there must be an accounting system to register the collections. Non-payment was avoided by careful recording of each individual's arrears and, ultimately, withdrawal of service from users in severe arrears; but the level of arrears nevertheless caused frequent cash flow crises (Abernethy et al., 2000).

How will river-basin organizations minimize these linked problems of compliance, delayed payment, and collection cost? The answer to this seems to be (as for some of the other issues raised above) that the separation of regulatory from service-delivery functions should substantially reduce this problem. The service deliverer must have a direct relationship with the water users, and indeed the trend towards user-controlled service organizations assists this. The regulatory organization on the other hand has different duties, and should collect its fees from a few major sources, principally the service-delivery organizations, but also including any others to which it grants permission for abstractions, pollution permits, or other water-related activities. On the whole, passive modes of collection may be sufficient for this.

5.8 Control of Expenditure

As river-basin organizations come into existence in an increasing number of countries, we will face another kind of issue: how will their expenditure levels be controlled?

These organizations should become, as far as possible, responsive to the interests of their own stake-holders. But the stake-holders are very diverse. Everybody is a water user; and most of us are water users in several different modes. Some may want new storage facilities to be built; others would prefer that costs be kept down as far as possible. The interests of birds, fish and other wildlife have to be accommodated somehow, along with other non-economic aspects of water, such as landscape beauty, waterfalls and the like. All of these things tend to have cost connotations in some way.

These matters cannot be satisfactorily resolved by creating river-basin organizations that are firmly embedded in governmental bureaucracy. A different and more responsive kind of organization is needed, which will be accountable to some council in which all principal stake-holders have a voice.

In these conditions, control of expenditure can be done transparently, with budgets approved in advance not by finance ministries, but by the people of the basin who will have to bear most of the costs and receive the consequent benefits.

The problems of expenditure control provide a strong reason why, at least in the initial stages, new river-basin organizations should not be asked to develop functions that are too large. Stake-holder councils and other similar institutions are likely to be rather meaningless and ineffective if they do not have the general right to formulate an annual budget, or at least to modify and approve (or reject) budget plans proposed to them by some part of the government bureaucracy. On the other hand, it is likely that a considerable time will pass before such councils are trusted to formulate and supervise a large capital budget effectively. Risks of incompetent or dishonest behaviour obviously exist. Capital expenditures also may frequently have impacts or benefits outside the basin where the works are constructed, so their financial management entails more complex procedures.

For such reasons it seems that it would be better to move along the path of developing the financing of regulatory functions first, which can be done within the basin and should be possible within the financial resources that can be generated from the kinds of funding sources described earlier.

To ensure public confidence in the management of such funds, we should expect that river-basin organizations will submit to some process of public audit, according to whatever the relevant national systems may be, perhaps sharing the same processes as are applied to (for example) public auditing of the accounts of local government bodies.

5.9 Impacts of Charging

The Economist (25 March 2000: 84), reporting after the World Water Forum, said,

> whether it is Australia or Rajasthan, once people understand the true cost of water services, they will conserve water, and even help to dig ditches if necessary. In return, they will want transparent prices and better service from both governments and private firms.

That is a clear statement of the standard economic view of the impacts of water charges: consumption will reduce, capital costs will be partially taken up by users, and customer pressures will cause organizational behaviour to improve. Is it true?

It is quite difficult to reconcile this optimistic view of the power of economics, with the findings from the detailed work of PMI-Niger (Abernethy *et al.*, 2000). There, in a country that is at or near the bottom of the *per capita* wealth scale and other human development scales, irrigation service fees are among the highest in the world. On the whole, fee collection rates are high. Farmers pay 20% of their gross crop value in fees. If the foregoing quotation is true, water should be used very efficiently in these circumstances. But the water productivity was found to be equivalent only to 20 US cents/m³, in terms of gross product value, at purchasing power parity (less than 5 US cents/m³, at nominal bank exchange rate). No signs of reduction in water consumption could be detected over ten seasons of monitoring.

There seem to be three sources of the difference between these observations and the view quoted earlier. First, *The Economist* was drawing lessons primarily from urban cases. Secondly, water is only one input to a production process, whether in agriculture or industry, and it is generally not a replaceable input; so if the user thinks that more water is necessary in order to realize the benefits of other inputs, that user will probably apply the extra water. But it seems likely that the third reason is the most influential. This is that farmers in Niger, as in most other developing countries, do not pay for the quantity of water they use. They pay heavily, but the charge is area-based, so the marginal cost of taking more water is zero. In the terms of the above quotation, the irrigation users may not come to 'understand the true cost of water services'. When the marginal price paid for additional supply is zero, the price system sends no signal to the users concerning the true cost of their consumption. Yet area-based irrigation charges, which produce this ineffective result, are the most usual kind in developing countries.

Urban users, whose quantitative needs are smaller and more measurable, normally are not in that situation. For them, the marginal cost of increasing their usage of water may be quite high.

This problem, that the major users of water (farmers) have no direct incentive to reduce their consumption, is not likely to change in the near future. Although we can measure pipe flow volumes acceptably, and the equipment cost for doing so is quite tolerable, devices for measuring flow volumes (as distinct from flow rates) in open channels are not available at the scales and costs required for the small land units typical of developing countries, especially in Asia. So the impacts of charging in metered urban systems can be quite different from those in small-holder irrigation.

One proposed approach to this problem is by charging, not to individuals, but to groups, for example to all the farmers along a single common channel. There is as yet little evidence that this is effective. At the level of the individual, it does not alter the incentive much. If there are 50 farmers sharing a metered source, each may calculate that, by taking an extra cubic metre of water he or she will obtain all the benefit of using it, but will pay only one-50th of its cost.

For a financial system to have a strong impact on water abstractions, it must also be designed to give incentive to the service-delivery organizations to reduce the conveyance losses in their systems. Both urban water-supply organizations and irrigation departments have until recently shown poor records and lack of concern about reducing losses. This is another area where separation of river-basin management from service delivery is helpful.

If the service-delivery organization has to buy the right to abstract water, it will be more strongly motivated to ensure that as much of that water as possible reaches a customer who pays for it. Water leaking from a canal or a pipe, in that system, means a direct financial loss to the service-delivery organization. Organizational separation also makes it easier to include in the financing arrangements some reflection of the value of leakage water that can be recovered by pumping from aquifers, which varies greatly according to factors like location, quality and aquifer depth.

There is also the problem of equity. As we move more towards the principle of payment for water services, can we feel sure that the poorer sections of society will have adequate access to water? Briscoe (1997: 349) put this problem clearly:

> The inequities of existing command-and-control mechanisms for water allocation in irrigated agriculture have been widely documented . . . Because water has rarely been formally managed as an economic good in developing countries, however, there is little information on the equity effects of a market-oriented management system.

The problems of inequity are particularly acute in urban water supply systems. One of the results of rapid urbanization in modern times has been that cities have grown much faster than the administrative and financial capacities for providing them with comprehensive public utilities. In many cities, therefore, piped water is now provided

in only a minor proportion of households, and many of the other households, unless they can install their own wells, must rely on small water vendors. Bhatia and Falkenmark (1992) gave data on the ratio between prices per litre charged by such vendors, and prices charged by public utilities, in a large range of cities. Their data show that, on average, the vendors' prices were about 16 times greater than the prices charged by the public organizations; and this ratio ranged up to 100 in very poor environments such as Port-au-Prince (Haiti) and Nouakchott (Mauritania). We may guess too that the water supplied by the vendors is generally of substantially lower quality.

This kind of information is not evidence that a 'market-oriented management system' will have necessarily bad effects for the poor. The vendors may not be over-charging: they have costs to meet and transport and labour to arrange, as well as the primary cost of the water. These enormous price ratios rather show the scale of under-investment in extending the public piped water systems. The under-investment in turn may be linked to inadequate prices charged for the products of those systems.

We can, however, feel relatively confident that there are better chances for restoring equity, under river-basin management organizations, than under the present systems of management. Traditional water rights have been rapidly eroded by the political and economic changes of the past two or three decades, and relatively few countries have succeeded in supporting the traditional systems, or in replacing them by modern systems based on water rights that are legally enforceable by their users. Basin management offers a way of redressing this situation, either through rights or licences. It would seem reasonable to accept the need for some payment, or increase of existing payments, in return for a better guarantee of supply or abstraction rights.

However we should also note the need for good, transparent public information programmes when such a policy change is under consideration. If public opinion is not prepared for changes of traditional patterns, and not informed about the benefits that they are intended to bring, they are likely to be rejected.

Charges, or increases of charges, are never going to be popular. It is futile to hope for that. However sound our economic logic, however much we may feel that a charge system can reduce distributional inequities, or improve water-use efficiencies, there will not be demonstrators in the streets demanding the introduction of such charges. It is good to keep this point in mind, as we think about possible beneficial impacts of charging policies. On the contrary, if charging systems are introduced where there were none before, the idea will almost certainly provoke a strong adverse reaction, and it must therefore be preceded by transparent explanations to the people about the benefits that are intended to be derived from the policy.

Yet there are strong arguments for saying that the interests of poor people can be better protected under a stronger system of financial management, which can be combined with high charges for heavy uses, financial penalties for pollution and other types of social costs, stabilizing of access and use rights, control of ground-water depletion, and a number of other desirable outcomes which at present countries struggle to achieve. The interests of the poor can be protected to the extent of some cross-subsidisation, introducing minimal prices for small domestic basic-needs quantities, or even as far as the zero rate for the first 6 m^3/person/month which is proposed in South Africa.

We should note however that cross-subsidization (supplying to some customers at prices that are below true costs, or even zero, and balancing the accounts by charging prices above true costs to others) is not in the long-term genuine equity, although it may be justified occasionally in special circumstances such as those South Africa faces. Non-equitable policies of this sort need to gain the support of a consensus of the water users of a basin, just as much as other water management policies.

The equity question can be regarded as yet another argument for keeping a separation between the regulatory and service-delivery functions. Research on equity effects will probably continue to be necessary for quite a long time. River-basin organizations should take some responsibility for monitoring these effects, and for encouraging the necessary research. They should be in the position to adjust their regulations, and the constraints on the service-delivery organizations, so as to take account of the need to limit the degree of inequity that may exist, and particularly to ensure access for all up to a certain basic level.

5.10 Conclusions, and a Possible Way Forward

River-basin organizations offer a promising way towards better and more equitable management of water resources. They need to be adequately financed, and it is better that their finance should be generated locally from among the users of their services, who should also have an effective voice in influencing their policies, than that their finance and policy should be determined centrally.

The ways of generating sufficient finance, in the case of poorer countries, are not yet certain, because of the weak financial situation of agricultural users, who account for the overwhelming majority of water consumption in most countries. The lessons that can be learned from the financing systems of richer countries are of limited relevance because of the different balance of user types.

The financing system has a strong effect on the behaviour of an organization, so the financing system for a new river-basin organization should be designed not only on accounting principles, but also with a clear view of the behaviour we want. The desirable characteristics of a river-basin organization could be summarized as follows:

- Its primary goal is to provide a framework of legality and security for all the uses of water, including environmental as well as human uses.
- A river-basin organization should have no function of delivering water

supplies or services, since one of its tasks will be to monitor and control the activities of service-providers.

- Its primary funding mechanism should be through licences for abstractions and disposals of water, supported to some extent by penalties for pollution, excessive use and other unwanted behaviour, and permits for non-abstractive uses.
- Its charging structure should reflect the scarcity and quality of the basin's water.
- Its charges should be structured to ensure that basic human needs can be satisfied cheaply.

The *possible* impacts of water charges on the behaviour of water users include these:

- Consumption reduces;
- Water-saving behaviour increases;
- Organizations perform better;
- Users contribute to the cost of capital works;
- Equity of access and of use changes.

As we have seen, the *actual* impacts depend significantly on the precise mode of application of the charging system, which must be designed carefully or it may not yield the desired outcomes.

Current systems of charging for water services have many defects. Irrigation charges are usually area-based, not volume-based, so they give no incentive for water saving, neither for the service-delivery organization to reduce leakages nor for the end-user to improve application efficiency. Charging rates are usually calculated on the basis of the cost of providing water delivery service, and sustaining a supply organization, but do not often reflect the scarcity or abundance of the water resource, or the quality of the water. Often economic logic is reversed, as poor city-dwellers pay more for low-quality domestic water, and farmers pay a marginal price of zero for water applications in excess of real need.

Systems of abstraction licences may be the most easily implemented method of addressing simultaneously these various problems, especially:

- to sustain the kind of river-basin organization that is needed;
- to give firm legality to long-standing traditional users;
- to protect principles of equity during rapid socio-economic changes; and
- to make possible more flexible systems of charging that will reflect scarcity and quality, and will follow some progressive scale so that basic needs (both personal water use and food production) can be satisfied at rates that are less than those charged for levels of use which exceed the fundamental human requirements.

Figures 5.1 and 3.2 represent the kind of pattern that is implied by these arguments. Figure 5.1 is what we may call a 'traditional system' in many developing countries. A number of service-providing agencies, which each have a basic single service function (irrigation department; electricity authority; water supply board, and so on) exist, independent of each other and each under the patronage of a ministry. (There may also be large industrial companies in the private sector, which are allowed to perform 'self-service', abstracting, using and disposing of water for their own functions; there may also be companies which are privatized water-service providers.) The service-providing entities are financed by some combination of government subsidies flowing down from the parent ministry and service-related fees flowing up from the users. The ratio between these downward and upward components varies among countries, and is changing over time: 20 years ago, the upward flows were generally much smaller than now, and the downward subsidy flows were much greater.

The pattern shown in Fig. 5.1 does not lend itself easily to integrated management, as many countries have found. The agencies and their patron ministries tend to show rivalry rather than co-operation. Without integrating the actions of these service-providing agencies, there can be little prospect of bringing about coherent management and socially acceptable principles of allocation of the water resource, and disposal of

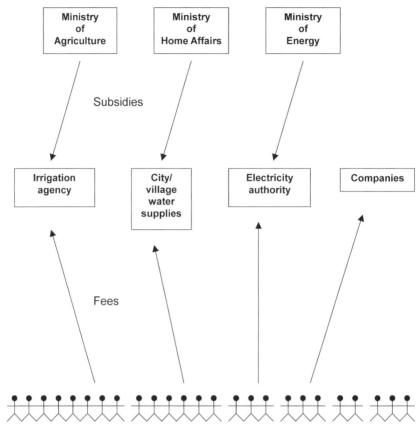

Fig. 5.1.   A common pattern of financial flows.

it after use. Yet it is extremely difficult to envisage dismantling these long-established structures, and it is not at all sure that the disruption involved would be beneficial.

Figure 5.2 indicates a 'new' system, leading towards a gradual integration. The existing system of organizations and financial flows can be left in place. Alongside it is the river-basin organization, which is given authority by the state to issue licences for water abstractions and disposals, and to charge appropriately for such licences. Initially the licence pattern will no doubt correspond closely to the existing pattern of uses. The new system however changes the processes by which the existing situation will evolve in future, and makes it more likely that the allocations of water can be brought into balance with future needs and with environmental goals.

The financial flows to the new organization would, in this model, come initially from charges on the service-providing organizations, including 'self-serving' industries and irrigation schemes. The flows would be expected to expand at a relatively early stage to include some flows from individuals, for example fees for licences to install new ground-water wells, and gradually to broaden their revenue sources by taking up some of the other financing options mentioned earlier, but leaving the primary revenue that can be derived from user fees to be collected by service-providers.

To ensure that this system responds to the general goals of the community, great care is needed in the design of its organization, its procedures of consultation, democracy, equity and so forth. Those important features are beyond the scope of this chapter;

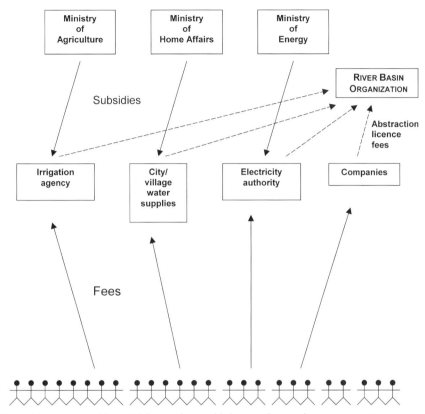

Fig. 5.2. A future pattern of financial flows, after establishment of a river-basin organization.

but the need for harmonizing the financial procedures with those social dimensions is clear.

References

Abernethy, C.L., Sally, H., Lonsway, K. and Maman, C. (2000) *Farmer-based Financing of Operations in the Niger Valley Irrigation Schemes.* Research Report 37, International Water Management Institute, Colombo, Sri Lanka.

Bhatia, R. and Falkenmark, M. (1992) Water resource policies and the urban poor : innovative approaches and policy imperatives. *International Conference on Water and Environment, Dublin, Ireland.*

Briscoe, J. (1997) Managing water as an economic good. In: Kay, M., Franks, T. and Smith, L. (eds) *Water: Economics, Management and Demand.* International Commission for Irrigation and Drainage/E. and F. N. Spon, London.

Briscoe, J. (1999) The changing face of water infrastructure financing in developing countries. *Water Resources Development* 15, 301–308.

Buckland, J. and Zabel, T. (1998) Economic and financial aspects of water management policies. In: Correia, F.N. (ed.) *Selected Issues in Water Resources Management in Europe,* Vol. 2. Balkema, Dordrecht, pp. 261–317.

Clarke, R. (1993) *Water: the International Crisis.* Earthscan, London.

Gleick, P.H. (1996). Meeting basic human needs for water. *Stockholm Water Front* 1, 6–7.

International Conference on Water and Environment (1992) *The Dublin Statement.* World Meteorological Office, Geneva.

Oorthuizen, J. and Kloezen, W.H. (1995) The other side of the coin: a case study of the impact of financial autonomy on irrigation management performance in the Philippines. *Irrigation and Drainage Systems* 9, 15–37.

Projet Management de l'Irrigation – Burkina Faso (1997). Analyse diagnostic et performances de cinq périmètres irriguées autour de barrages au Burkina Faso. PMI – BF and International Irrigation Management Institute, Ouagadougou.

Schiffler, M., Köppen, H., Lohmann, R., Schmidt, A., Wächter, A. and Widmann, C. (1994) *Water Demand Management in an Arid Country: the Case of Jordan with Special Reference to Industry.* German Development Institute, Berlin.

Seckler, D., de Silva, R. and Amarasinghe, U. (1996) The IIMI indicator of international water scarcity. In: Richter, J., Wolff, P., Franzen, H. and Heim, F. (eds) *Strategies for Intersectoral Water Management in Developing Countries: Challenges and Consequences for Agriculture.* DSE/ZEL, Feldafing/Zschortau, Germany.

Svendsen, M. (1992). *Assessing Effects of Policy Change on Philippine Irrigation Performance.* Irrigation Performance Paper 2, International Food Policy Research Institute (IFPRI), Washington DC.

Tardieu, H. (2001). Water management for irrigation and environment in a water stressed basin in south-west France: charging is an important tool, but is it sufficient? In: Abernethy, C.L. (ed.) *Intersectoral Management of River Basins.* IWMI, Colombo, Sri Lanka.

World Bank (1992). *Development and the Environment: World Development Report.* Oxford University Press, New York.

World Bank (1993). *Water Resources Management: a World Bank Policy Paper.* World Bank, Washington, DC.

World Bank (1999). *Knowledge for Development: World Development Report, 1998–99.* Oxford University Press, New York.

6 Water Management for Irrigation and Environment in a Water-stressed Basin in Southwest France

Henri Tardieu

6.1 Introduction

In France, the water management issue is no longer seen as a question of developing stakeholders' participation or transferring State competence to user associations. As for the other countries with a complete and complex institutional framework, the challenge now is to define clearly the role of each water management stakeholder and to answer two remaining questions:

- How do we ensure sustainability of the investments by raising the price of water without discouraging economic development?
- How do we share water among users when resources are scarce?

The general answer to these questions relies on the two basic principles of good water management leading to sustainable development:

- Since it consumes more than 70% of the available water during low flow periods, irrigated agriculture must respect the other uses by limiting its demand to the allocated volume;
- Since it involves large and long-term public investments, irrigated agriculture must bear at least the 'sustainability cost' (Tardieu and Préfol, 2002) of the upstream water resources.

Such a general answer is of course largely case-specific and should be adjusted to each institutional framework.

France, like other Euro-Mediterranean countries, has a long history of water development, born from water scarcity and a constant search for the best agricultural use and the fairest sharing. A complex institutional structure has progressively been set up to develop private initiatives within a public service framework.

During the last century, Authorized User Associations (ASAs) were developed. They were formal institutions constituted by landowners for sharing the construction and management of irrigation systems. In the 1950s, the State created, within a more ambitious land-use planning framework, Regional Development Companies (SARs), which are public corporations with a 'concession' from the State, to develop water resources and manage irrigation schemes in the southern regions of France. Well subsidized by the State at the beginning, the SARs now cover their costs from the payments of their customers. This management is now financially sustainable as it includes the provisions necessary to maintain the investments under concession. It nevertheless keeps the basic characteristics of a French public service: continuity, equity, sustainability and transparency.

©CAB International 2005. *Irrigation and River Basin Management*
(ed. M. Svendsen)

Finally, Basin Organizations were set up more recently, with a widened mandate to include management and protection of the environment and seek a global consensus on water management by using dialogue and financial incentives, while the State keeps the role of regulation.

After a short discussion about the stakeholders in French irrigation and water management, this paper addresses both socioeconomic questions stated above, with a specific discussion of the case of the Neste system, a water-stressed basin in the southwest of France.

6.2 The Key Stakeholders in Irrigation and Water Management in France

6.2.1 The individual level: farmers

Farmers aim to satisfy the objectives they select for their household (to ensure a minimum revenue), their enterprise (to maximize profits, to minimize risks and to improve the quality of the products) and their land (to be sustainable). Each one freely chooses the crops to grow on the basis of advice from his profession with due consideration given to the market. He consequently optimizes the management of the production factors, including the on-farm irrigation system.

The value of water in irrigated agriculture varies, largely due to the heterogeneity of the production systems. The cost of irrigation water is generally relatively high in the Mediterranean regions, implying high performances with high value-added crops.

The constraints of agricultural competitiveness make the irrigator very sensitive to the reliability of water supply and, of course, to its cost. For each crop in a given cropping pattern, water value can be assessed. Water demand can thus be represented by a graph of water value per unit volume. Such a graph suggests how an irrigator reacts as the water price varies.

6.2.2 The small system level: ASAs

Gathering irrigators through an association that owns and/or manages common assets is the first and the oldest way to manage collective irrigation (Lesbats *et al.*, 1996). Such associations bring together the landowners concerned with the irrigation system. They are self-managed organizations, and, in France, are based on a legal framework developed in the 19th century. This framework provides all the authority needed to manage the irrigation system.

The statutes of ASAs are public and confer the capacity to act for the public good, particularly in the matter of cost recovery where they follow the rules of public accounting. Costs are shared in proportion to the involvement of each owner in the project area, generally as a function of his irrigated area.

These associations have a very long lifetime, since the properties are irrevocably engaged in the association. This long experience provides valuable lessons. Initially, the collective participation of members is exemplary. They define their projects according to their needs and their means of fulfilling them. They personally ensure operation and minor maintenance. The apparent cost, corresponding only to monetary expenditure, is thus largely below the comprehensive cost of water.

However, it sometimes happens that the necessary solidarity decreases and the ASA goes wrong by lack of involvement and lack of professionalism. Members are then concerned about the immediate balance of the accounts, and cut down the maintenance expenditure. This entails serious consequences in terms of quality and continuity of service.

To preserve or re-establish the durability of their systems, at the conception as well as at the management stage, ASAs have currently been calling for the assistance of the State. The ASA statutes indeed foresee that in case of bankruptcy of the ASA, the State representative must replace the ASA President. Considering the State's other involvements, ASAs are now looking for professional advice, particularly from the SARs (see below).

Such a complementary relationship between ASAs and a technically competent body can be organized to guarantee sustainable management. This is the case with the design and/or maintenance contracts offered by Compagnie d'Aménagement des Coteaux de Gascogne (CACG), one of the SARs.

6.2.3 The large systems level: SARs

Created about 40 years ago in the southern regions of France where water was a limiting factor to development, the SARs are characterized by the originality of their mission and statutes (Plantey *et al.*, 1996).

Their mission, defined by the concession contract with the State, deals with the implementation and operation of hydraulic projects necessary for the development of their region. Managing the conceded water resources, they ensure the conveyance of water to the centres of urban and industrial consumption, and the distribution of water in rural irrigated areas. For this purpose, they have broad rights and obligations as 'owners' of the works, but without the right to sell them.

Their statutes are similar to those of private companies, implying rules of sound management and economic efficiency. The majority of the shareholders are public, and so is the governance. Local authorities (Départements[1] and Regions) therefore have control of the strategic resource in the name of the public good for all water users. The agricultural users are represented on the Board and participate in governance as the SARs' private shareholders.

In accordance with the terms of their concession contracts, the action of the SARs is guided by the principles of sustainable management of a public service.

- Quality and continuity of water service;
- Equity when water is to be shared between users;
- Sustainability with adequate provisions for long-term maintenance;
- Transparency of the management and accountability to the Board.

The SARs have been relatively successful at balancing resources and needs through the practice of integrated water management. Despite the relative scarcity of the resource in the French Mediterranean regions, water shortages and conflicts among users are no longer a major concern in the systems managed by the SARs (Tardieu and Plantey, 1999).

When a crisis situation arises, high-tech equipment and tested methods of water sharing allow equitable management of the resource. This is typically the case for the Neste system managed by CACG.

6.2.4 The large catchment level

A Basin Committee, a sort of water parliament where users, local authorities and government are represented, is in charge of conserving the water environment and of water management policy in one of the six French large catchment basins. It develops, in collaboration with the State administration, the long-term water policy plan.

Water Agencies are the executive bodies of Basin Committees. Taxes, collected by the Water Agencies in accordance with decisions of the Basin Committee, discourage polluters and consumers from polluting and consuming. This incentive to behave in a more responsible manner is coupled with a financial policy since the revenue from the taxes is allocated to financial assistance for pollution abatement and for conservation and development of water resources. For irrigation particularly, Water Agencies

---

[1]   A Département is a local government unit, with population usually in the order of 500,000.

contribute to investments in modernization and regulation, which are important sources of water savings.

After 30 years, this system based on the principles of solidarity and equity (the polluter pays and the consumer pays) is well accepted by the public. It should be understood that the French Water Agencies do not have direct responsibilities in water system management, unlike the bodies described above.

6.2.5 The State level

According to the terms of the 1992 Water Law, it is not the State's responsibility to ensure directly the operational management of water resources, except for very large rivers. Its authority should guarantee the respect of the necessary regulation of water uses, which are subject to previous authorization. Elaborating and updating the rules should be carried out in consensus with the members of the water community to minimize the number of rule-breakers.

Finally, the State is the owner of large hydraulic works used for irrigation purposes and delegated to the SARs by concession contract. As a consequence, it supervises both the maintenance and the best use of the assets in order to meet all water demands.

Although this presentation of the French institutional framework in water management is highly simplified, it does help to clarify the respective roles of different stakeholders in different areas.

- Basin planning and financial policy: the Basin Committee seeks a consensus to reconcile all users, both among themselves and with the environment, in a global approach to water management using financial incentives.
- Operational management: the SARs manage water resources by contracting with users, and ensure the sustainability of the assets. The ASAs have almost the same objectives but for smaller irrigation systems.

- Regulation and law enforcement: the State sets up regulatory measures, keeping in mind both the need for consensus and acceptance and its own capability for applying them to all users.

The French organizational set-up for water is characterized by a 'public/private' mix. The freedom of the private initiative is balanced by the responsibility for the public good. The economic efficiency of private management is associated with the sustainability of the public service.

6.3 The Neste System: an Example of Controlled Water Management

Water management in the Neste system provides an important example of successful application of a set of rules, consultation methods and high-tech controls, but also represents a system facing a regulation problem because water supply is currently exceeded by demand. The essential features of the water management agency, the CACG, that make success of 'controlled water management' possible include: (i) the institutional originality of the SARs with a public mission and private management; (ii) their joint experience in regional development and water management; (iii) their capacity to carry out maintenance and asset conservation; and (iv) the practices they follow in pricing water in combination with a quota system. Economic analysis of the water value in each use, particularly for irrigation, may clarify the allocation of water and validate the regulation tools used in 'controlled water management'.

6.3.1 Presentation of the Neste system

In 1990, after a serious crisis involving conflicts between users due to water scarcity, a new management system was established (Tardieu, 1991). In operation for 10 years, it can be described as follows, with its successes and limitations.

6.3.1.1 Location

- A 10,000-km^2 basin located in the southwest of France with 650 mm of rainfall, where irrigation is necessary for most kinds of agricultural production (Fig. 6.1);
- Surface water is the only resource for urban and industrial uses because of

lack of groundwater, and recharged rivers (1300 km) are the common resource for every user.

6.3.1.2 Water users

- Fish, wildlife and tourism need 250 M m^3/year to augment low flows;[2]

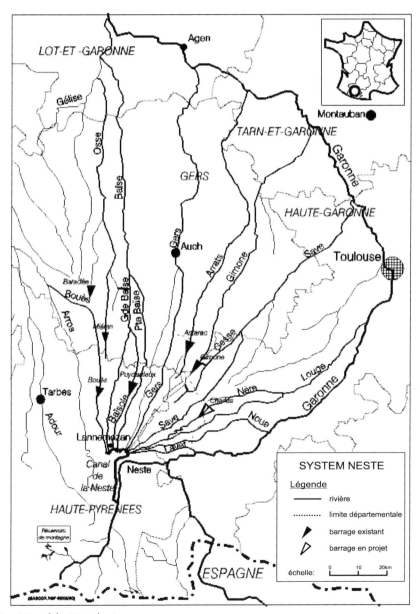

Fig. 6.1. Map of the Neste basin.

- 200,000 inhabitants consume 13 M m³/year;
- 51,000 irrigated ha (28,000 l/s subscribed by 3000 irrigators) consume (average) 70–95 M m³/year (dry years);
- A 10,000-ha waiting list for irrigation contracts (equivalent to 6000 l/s) exists.

6.3.1.3 Water resources

- The Neste Canal (a State concession to CACG) diverts 250 M m³ of the natural flow of the river Neste;
- Stored resources contribute 100 M m³, of which 48 M m³ are stored in mountain hydroelectric reservoirs and 52 M m³ in CACG lakes (also State concessions).

6.3.1.4 Types of withdrawal

- Individual withdrawals (14,500 l/s) subscribed through 'conventions de restitution', or 'pour-back contracts';
- Collective withdrawals by ASAs or CACG irrigation networks (13,500 l/s).

6.3.1.5 Monitoring and remote control

- Supply monitored through 200 river flow meters, 40 dam and canal gates, and 150 pumping stations under remote control;
- Demand monitored through 1500 individual water meters (checked three to four times a year), 6000 meters on collective networks, and 150 pumping stations under continuous monitoring.

6.3.1.6 Supply/demand balancing

- Currently the balance is ensured with a failure rate of 1 year in 4;
- For a more comfortable balance, an additional resource of 43 M m³ is needed for the waiting list and 7 M m³ to reduce the failure rate.

6.3.2 Management rules: the contract, individual and collective

Each user signs with CACG a water contract called *convention de restitution* guaranteeing that his/her withdrawal is balanced out by an equivalent upstream replenishment. The contract states a maximum diversion flow and a subscribed volume, or *quota*, with a two-tiered price. The first price is a function of the allowed flow (€50 per l/s). The second price, the *over-consumption* price, is a function of the volume consumed above the quota. (€0.10/m³ above the 4000-m³ quota).

Thus, there are two limits on the abstraction of water by the user: a flow rate limit, and a volume limit. If the authorized flow rate per hectare is 1 l/s and the volume quota for the year is 4000 m³, the user may, in effect, abstract water for 1110 h before the quota is exceeded. The extra payment required from those who exceed their volume quota is €0.10/m³. The price step is thus large. By paying €50, the user becomes entitled to take up to 4000 m³, so if the full quota is taken its average price per m³ is €0.012. If the quota is exceeded by the user, the marginal price for taking more water rises immediately to €0.10, almost eight times the base rate.

In reality, the user will often take less than the quota, particularly during rainy years. In that case the payment of €50 remains, so in effect the average price paid per cubic meter is more than €0.012, which is the minimum possible. Over a long period, the average price actually paid is close to €0.02 per m³.

The contract also fixes penalties for the user (in case of withdrawing above the allowed flow rate or the lack of a water meter) and for CACG (in case of quota reduction). As demand exceeds supply even when dams are full, the Neste Commission, which brings together all water stakeholders from the five Départements involved, decided to start a waiting list of applicants. All applications who are rejected because of supply

---

2   M m³ = million (10⁶) cubic metres.

limitations are registered in an open access file which totals 6,000 l/s to date. Newly created resources and contract terminations allow a few of these filed applications to be accepted annually, according to priority rules (young farmers) or to seniority on the waiting list. All the yearly contracts get a collective withdrawal authorization in each of the five Départements.

When dams are not full or when it is anticipated that the Neste river flow will decrease, the Neste Commission meets before the irrigation season to decide on a quota reduction. The choice of the meeting date is the result of a compromise between the possibility of making a sound hydrological forecast and the possibility for farmers to adjust cropping patterns or inputs.

During the irrigation period, water meters are checked regularly. If a quota seems likely to be exceeded, a warning letter is sent by CACG to the irrigator. Quota overruns are billed at the end of the season. CACG is also in charge of resource management, which it carries out through computerized remote control (RIO software), tactical water management in order to save water transferred from the remote-controlled dams (checked every 3 h) and strategic water management in order to optimize water allocation between irrigation and river wildlife. The objective in the latter case is to empty the reservoirs by the end of the low flow period with a failure rate of one in ten (weekly check).

After the irrigation season, water management performance is assessed in terms of meeting or exceeding minimum wildlife flows, supply of volumes subscribed by irrigators and water savings throughout the system.

- Since 1990, when the system was first put in operation, failures to maintain wildlife minimum flows have been rare, amounting to only 1–2 days/year over a reach of a few kilometres, as compared to the drying up of several dozens of kilometres over several weeks in 1989.
- However, irrigators' quotas have been reduced 4 years in 10 (though one of

these reductions was later cancelled), and in 3 years out of 10, users have been prevented from exceeding their quotas regardless of extra payments. More extreme State intervention (by the Prefect at the Département level), which constitutes the ultimate recourse when the crisis cannot be met by means of quotas, has not taken place.

- After using the RIO software for 10 years, experience has shown water savings of more than 20% of the managed volume.

6.3.3 Successes and problems of the system: can a limited supply be regulated?

In the Neste system, both principles of good water management are respected:

- Water is shared in such a way that fish and wildlife are preserved all along the 1300 km of controlled rivers and irrigators receive their contracted volumes;
- CACG, on behalf of the 'conceding' State, bills users for the cost of the service and gets the financial means to cover at least the system's 'sustainability cost' assuring the maintenance of assets worth €540 million at current prices.

There has clearly been progress in comparison with the two 'wrong practices' of the previous period: (i) daily interventions by the Prefect which irrigators circumvented by over-investing in pumping capacities; and (ii) the inability to charge for the 'resource' part of the water service, thus leading to asset jeopardy.

One direct positive consequence of the new arrangements is that irrigators are driven to save water and optimize their cropping patterns through a system of sound and sustainable incentives far more valuable than any media campaign. It also induces a renewed interest on the part of research and extension networks in quota optimization through revised cropping patterns and input use (Balas and Deumier, 1993).

However, a fundamental question remains, which concerns continued spatial development: what about the waiting list's demand if water resource creation is hampered by procedural problems and limited public funds? One solution sometimes envisaged is to reduce the quotas of current irrigators in order to admit the new applicants. This bad solution, which includes issues of equity, economic efficiency, social acceptance and technical and agricultural management, has been discussed in a separate paper (Tardieu, 1999) and is summarized in Section 6.6 below.

6.4 Key Actors and Essential Basin Functions

In addition to the key actors described in the first section of this chapter, other government units also play important roles in basin management. These include towns, the Ministry of the Environment and the Ministry of Agriculture. The interaction between key actors and essential basin functions is shown in Table 6.1. It should be noted that almost all of the water used in the basin comes from surface water sources since little groundwater is available, and that surface water is used primarily for domestic, agricultural and environmental purposes.

The point that stands out most sharply from a review of Table 6.1 is the central role played by the SARs in the French model. In the Neste, CACG plays major roles in constructing and maintaining facilities and in allocating and distributing water to users. The roles played by the irrigation associations are, by comparison, rather modest. Tools employed for allocating water are not solely administrative. The tiered pricing system for water deliveries used by CACG and withdrawal taxes levied by the Water Agency are important tools that help to determine the volumes of water that will be allocated to various users. The Ministry of Environment also plays an important role in providing the 'policing' powers to enforce allocation and use rules.

Another important feature of the matrix is the important planning and monitoring role played by the Basin Committee and the Water Agency, its executive arm. Because the Basin Committee includes most of the interested basin stakeholders, their involvement at the individual level is subsumed into the intersection between CACG and the planning function in the table.

The Ministry of Agriculture, which represents the State as the owner of the infrastructure, plays an important regulatory role in monitoring the care and maintenance given to the facilities by CACG, the concessionaire.

6.5 How to Charge for Irrigation Without Threatening Economic Development?

Whichever institutional framework one chooses, a fundamental challenge is to cover the full cost of water used by raising the water price and collecting fees from users. For most irrigation systems managed by government agencies, public subsidies are now limited by budget constraints. Such subsidies may consist of paying salaries of operating personnel, covering heavy maintenance or rehabilitation costs, or underpricing energy, etc. For us Europeans, the principle that 'users pay for water' will be the basis of the new European Water Directive. Some targeted and transparent subsidies will still be acceptable, but on the condition that they will be gradually phased out. This objective of an irrigation system breaking even on the basis of its own water charge revenue is not impossible to reach, and is already the case in several regions of France.

However, the economic and social consequences of water price rises can be serious, as shown by the following examples of likely risks:

- Overall reduction in the country's agricultural production, making it impossible to reach the goal frequently assigned to irrigation of securing food self-sufficiency. This consequence may

Table 6.1. Key actors and essential basin management functions in the Neste system basin.

Key actors	Surface water								Groundwater							Wastewater						Ag. returns			
	Plan (basin-level)	Allocate water[a]	Distribute water	Construct facilities	Maintain facilities	Monitor quality	Ensure quality	Protect against flooding	Plan (basin-level)	Allocate water	Withdraw/distribute water	Construct facilities	Maintain facilities	Monitor quality	Ensure quality	Plan (basin-level)	Construct facilities	Operate facilities	Maintain facilities	Monitor quality	Enforce quality	Construct facilities	Maintain facilities	Monitor quality	Enforce quality
Individual Irrigators	•		•	•							•	•	•									●	●		
Irrigation Associations			•	•	•																				
Towns							●	●			•	•	•				●	●	●						
CACG		●	●	●	●		●				•	•	•			•				•				•	
Agence de l'eau/Comité de Bassin	●					●			●							●				●				●	
Ministry of Environment	●	●				●		●	●	•				●	●	●					●				●
Ministry of Agriculture		•			●																			●	

●, Major role; •, minor role.

CACG, Compagnie d'Aménagement des Coteaux de Gascogne.

[a]And enforce quantity (Ministry of Environment).

be accepted if the country can maintain its 'food sovereignty' (FAO, 1996). A regular increase in the price of water has recently been started in Tunisia, except for cereals, for which water charges have been kept constant.

- Higher food prices for urban consumers, which induces larger food imports and some losses of internal market shares for irrigating farmers. This has already been verified in various African countries.
- Lower agricultural income, hence increased rural poverty and population migration towards towns. Even if the irrigating farmers are not the most vulnerable in economic terms – since they can use a wider range of crops – the economic development of rural populations must remain irrigation's fundamental objective.

On the other hand, the 'true prices' process can also generate some benefits:

- A new respect for water, which improves management efficiency;
- An incentive to choose the most profitable crops and to maximize comparative advantage;
- A means to know which assets have to be maintained, and which investments have to be made.

The price adjustment process has to be conducted with great care, taking into account the economic consequences on production. This is done by analysing the value of water for the farmer, i.e. the additional added value per water unit (m^3) provided by irrigated crops as compared with rainfed ones.

6.5.1 Full cost and 'sustainability cost'

Before tackling this issue, it is worth restating the definition of the full cost of water from the point of view of the agency responsible for water resource acquisition and distribution.

The *full cost* of water includes the following.

- Operating costs: staff, energy, daily upkeep;
- Investment-linked costs: depreciation/maintenance/renewal, financial costs of the initial investment.

A water price set at this level secures a balanced budget for the managing agency without any subsidies. In France, this price is about €0.15/m^3 for the large irrigation schemes, where water charges are based on the *full cost* of water with the first investment partially subsidized.

However, the cost of major headworks (reservoirs, transfer canals) is generally not included. The rationale for such undercharging is based on the consideration that these works are both strategic and multipurpose and that they were created for the sake of regional development at a time when economies were more State-backed and more protected. Today, countries where such infrastructure is paid for by water users instead of taxpayers are rare. Nevertheless, it is the objective that has been set for irrigation, notably in France, with a transition period allowing a smooth evolution of production systems.

During that phase, water charges are meant to cover what will be called the *sustainability cost* of water, something that, in the case of heavy, long-life investments, is very different from the *full cost*.

- *Sustainability cost* = operating cost + maintenance and sustainable renewal cost; or
- *Sustainability cost* = *full cost* − financial cost of initial investment.

With a water price set at the sustainability cost level, no new investment is possible; but budget constraints are met, and sustainable operation and maintenance ensured without having to resort to public funding.

Table 6.2 shows a simplified example of the costs generated by the water resource infrastructure, which, when added to the water distribution cost, yield the full cost of water supply. Values are drawn from the case of the Neste system, where actual annual costs of a storage reservoir feeding

Table 6.2. Sustainability cost and full cost in the Neste system.

Item	Cents€/ m³	Cents€/ m³
Operation and daily upkeep	0.75	
Maintenance/renewal (0.5% × investment cost)	0.75	
Subtotal: sustainability cost		1.5
Financial cost (long-term interest rate: 4.6%)	6.9	
Total: full cost		8.4

the river amount to about €1.5/m³ with a quasi-infinite reservoir lifetime.

A water price covering the *sustainability cost* of €0.015 m³ is socially acceptable and, after the public funding of the initial investment, prevents the need for further subsidies. In the transition phase, irrigation distribution costs are calculated at around €0.15/m³, i.e. at *full* cost pricing, while the irrigator's share of the water resource costs is charged more around €0.015/m³, i.e. at *sustainability* cost pricing.

6.5.2 Water strategic value

On the basis of the existing farming infrastructure, the *strategic value* (*Vs*) corresponds to the optimum combination between irrigated and non-irrigated crops, with a given cropping pattern:

$$Vs = \frac{VAI - VANI}{VI}$$

where VAI = value added from irrigated crops (before deducting the cost of water); VANI = value added from non-irrigated crops (rainfed crops) which could be cultivated instead of irrigated crops; and VI = volume of water allocated to irrigation.

This value[3] reflects strategic choices made by the farmer at a point in time when he can still modify his cropping pattern, and adjust his irrigation practice to a variable allocated water volume. It is the result of a decision taken once or twice a year and it should at least cover the cost of irrigation for it to be profitable.

Values calculated for each irrigated crop can then be plotted against the volumes of water required to produce the various crops, i.e. water demand curves showing the decreasing returns of irrigated crops (Fig. 6.2).

The following remarks can be made on the *Vs* formula presented above.

- Variations in crop prices (domestic or world price) can lead to changes in value that rule out irrigation or, on the contrary, that result in a stronger water demand. This is particularly the case for cereals, whose water valuation is relatively feeble but which call for large volumes of water.
- Changes in the yield or added value of a given rainfed crop can paradoxically

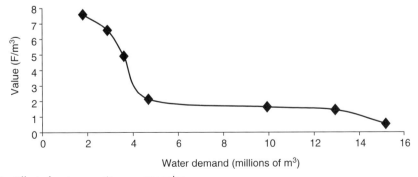

Fig. 6.2. Effect of water quantity on water value.

---

[3] See also Molden *et al.* (1998).

entail changes in irrigation water demand. For instance, a specific subsidy to rainfed durum ('hard' wheat) makes it an alternative to irrigated maize in the driest parts of southern France. On the other hand, the probable diminishing profitability of cattle breeding (as a result of the Common Agricultural Policy of the European Union) will increase water demand for irrigated cereals. The improvement of rainfed crops in Sahelian Africa, leading to lower import prices, may lead to reserving irrigation water for high added-value crops such as vegetables or fruits.

● Improving irrigation effectiveness, thus diminishing the formula's denominator, increases water value across the board.

6.5.3 Water's strategic value, price and budget constraints

By comparing the strategic value of water for the farmer and its cost, it is easy to know the average price that will balance the irrigation manager's budget. The problem for the irrigation agency, and for the State which backs it, is the following. Water price rises, which help to balance its budget, have a negative effect on water sales and, hence, a tendency to raise the costs of each cubic

metre sold, since irrigation costs are mostly fixed ones (depreciation, financing and maintenance costs). This is a vicious circle leading inevitably to the collapse of the system. That is why, in a now-transparent management environment, the State may find it advantageous to keep on financing intensification and modernization projects, thus boosting irrigated agriculture and increasing its own chances of recouping heavy sunk costs.

The concept of *sustainability cost* as described above is essential, for it constitutes the lowest price the State can accept (Fig. 6.3). If the water price does not cover the sustainability cost and exceeds the strategic value of water for farmers (for at least one given existing crop), this means that a long-term public subsidy through water charges will be necessary to maintain that irrigated crop in the country or region considered. The opening up of agricultural markets and the new transparency in world trade will make this practice impossible in the future.

On the other hand, it does not seem economically sound to dismantle entire sectors of irrigated agriculture on the principle that the *full cost* of water should be covered . . . at all costs! This would mean that today's irrigators would have to pay for investments which will also be used by future generations, which would be unfair, justifying a certain amount of public subsidy to help start the economic development process.

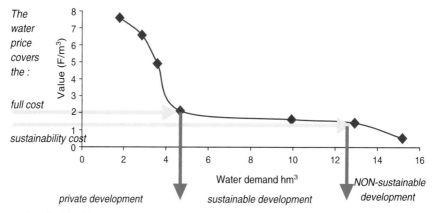

Fig. 6.3. Sustainable allocation of water.

So, when embarking on the true-price process in irrigation, it is essential to have a good understanding of *strategic value* of water to be able to derive demand curves by farm types and by region. Economic data on irrigated crops are not always available. This is one negative consequence of the disappearance of public irrigation agencies. The overall regulation of irrigation investments and of agricultural production requires governments to allocate some funds to data collection and processing that 'privatized' agencies can no longer afford. It should not be forgotten that economic issues to be documented should also include rainfed crops, to be in a position to correctly appraise water value by comparison between alternative agroeconomic systems.

6.6 Is Water Pricing Useful for Controlling Water Allocations?

The points made above assume that an essential prerequisite has been met – the clear identification of the economic agents who measure, buy, and sell irrigation water. This is an often heavy but always decisive task, which precedes and accompanies the true-price process in irrigation. It requires moving away from the idea that water is a free gift from the State towards the concept that irrigation is a service supplied by a provider to a client farmer. The critical question of how this transfer should be accomplished is the subject of many workshops. Let it just be said that, wherever water is scarce, it is very tempting to use the newly established economic links between 'supplier' and 'customer' to try to regulate water use through prices.

Indeed, after the beneficial disengagement of the State from direct management of irrigation schemes, some think that water allocation can also be taken care of solely by price mechanisms. To what extent is such price-based regulation reliable?

Water allocation regulation consists of inducing each economic agent to respect the volume of water allocated by the public authority. Is the pricing of water sufficient to avoid allocation crises in water-scarce systems? Can it settle intersectoral disputes between competing uses? Can it improve water distribution among farmers?

6.6.1 Quotas and pricing: instruments for allocation regulation

In many water-scarce regions, water quotas are allocated to farmers by sub-basin or by region. For the public authority, the problem is to ensure that these allocations are respected. The answer is usually of the law enforcement type, forbidding off-takes and penalizing those who withdraw excessive amounts. This type of regulation generates economic inefficiency and, sometimes, corruption. Therefore, it is highly tempting to use the price of water to avoid disputes between users, provided all participants have been identified and the service billed has been clearly defined.

From analysis of the *marginal value* of water,[4] a method of water pricing can be derived with a view to attempting such regulation. It will necessarily be step pricing, i.e. a discontinuous series of price levels, increasing with water demand. The higher price step, which will counteract the marginal value, must be higher than the lower price step, which itself is calculated to cover at least the sustainability cost of supplying the water and also to secure the farmer's income.

The fairly simple system set up in the Neste system consists of the following elements:

- An allocated *quota*, priced at a fixed amount which is the same whether the user takes it all or takes less;
- An *over-consumption* price for each unit of water above the quota.

---

[4]   Defined here as the value of the additional production that is brought about by one additional cubic metre of water applied during the irrigation period.

The overall volume *quota* must be compatible with the limited resources allocated to irrigation, as opposed to other competing uses. For a given existing irrigated area, there exists a corresponding volume *quota* per hectare, which has to be regulated with a price step high enough to deter over-consumption.

However, efficient regulation is based on understandable and practical water charges within a freely negotiated contract. Charges are useless if they cannot be recovered. Excessive *over-consumption* prices can only lead to jeopardized contracts and then to legal prosecutions, which is precisely the regulation mode to be avoided.

Experience from the Neste system shows that a price step between the *quota* price and the *over-consumption* price exceeding €0.10/m³ would not be socially acceptable at present. The approach outlined here provides useful strategic guidelines for fixing the volume *quota* per hectare. In consideration of the marginal value graph (Fig. 6.3), the highest price step clarifies the concept of socially acceptable minimum quota. If the *quota* is too far below the optimum needs (less than 80% of those needs) the system does not work in a dry year, and crises can be frequent, with stalled contracts and prosecution by the public authorities.

6.6.2 Best practices in water pricing and water resources development

Stemming from this discussion, three basic ideas can be emphasized:

- Water *step pricing* can help to regulate the allocation system if the *quota* and the *over-consumption* price are set in consideration of the marginal value of water and the social acceptability of the water charge.
- *Quotas* that are too low compared to the crop need cannot be regulated by pricing, and lead to economic inefficiency linked to enforcing inapplicable rules. The quest for equity at all costs in a system with limited resources leads to the same result.

- Increasing water resources in a water-scarce system makes it possible, over and above the direct economic benefits, to rebuild collective regulation based on a sound *quota* + price contract, which will leave each farmer free to manage his irrigation efficiently according to his own water value function.

Such a water pricing strategy, together with the investments needed to create new water resources, are necessary to help farmers face open market competition. Other requirements are a guaranteed and clearly contracted water supply and full supplier responsibility for irrigation management without State intervention.

Nevertheless, it is clear that such a system of price regulation can only work smoothly within a narrow range of economic variables – water price and water value – and it is the State's responsibility not only to identify this range but also to be ready to lay down rules on economically unacceptable behaviours. Only this type of strong State makes it possible for the managing agency to make efficient and economic use of price regulation.

The main advantage of such a framework is to give back to farmers the freedom to optimize their choice of crops and their management of irrigated or non-irrigated agriculture. This optimization becomes even more complex in the context of competitive world markets.

One prerequisite to the efficiency of this economic approach lies in the identification of the relevant agents (managing agency, individual farmers, water user groups), the clear and transparent content of their contract relations (water price, allocated water volume), and the capacity to measure the traded economic good (water meters). This indeed represents a move towards water markets (Kosciusko-Morizet *et al.*, 1998), but analysis of the value of irrigation water – particularly its *marginal value* – shows that it would be unwise to go further along this line, especially when it comes to free bidding for water *quotas*, given, on the one hand, the disproportion between the marginal value and the socially acceptable price of water and, on the other hand, the necessary equity in the sharing of a socially highly valued good.

6.7 Conclusion and Lessons Learned

The role of service-oriented organizations in irrigation and water management is now largely accepted as a prerequisite for implementing good control of water allocation and ensuring the sustainability of economic development (Malano and van Hofwegen, 1999). The transfer of management to water users' associations under control of an integrated basin authority is one possible solution, frequently described but too recent to be completely convincing. In fact, this type of solution often leaves unanswered the two important questions raised at the beginning of this paper, i.e. how to balance the budget by raising the water price and how to reach a fair sharing of scarce water among users. If water management is transferred to water users' associations, the process must be implemented carefully. The main idea is to promote 'self-management without abandonment' by transferring to socially strong users' associations responsibilities adapted to their capacity while keeping a tight partnership with a professional water manager.

The French history of irrigation management emphasizes the efficiency of some other solutions, such as management by SARs – mixed public-private companies linked to the Government by concession contracts. With the experience of such management, we can propose two recommendations in the very difficult debate on water pricing and allocation.

- Firstly, a cautious but firm move towards *sustainability cost pricing* is recommended. This involves charging the amount necessary to ensure the sustainability of the assets, i.e. operation, maintenance and renewal costs, but not the full financial cost of the initial investment or of the most recent rehabilitation. To correspond exactly to

sustainability, the price charged must cover all costs incurred in delivering each drop of water from the dam to the crop. At this level of cost recovery, there is no further need of current subsidies for staff, for repairs, for energy or for future rehabilitation. The subsidy's vicious circle is broken. Such development is sustainable, even though it is not designed to recover the initial investment.

- Secondly, a *step pricing* system based on water metering is recommended in order to facilitate control of allocation in a fair and transparent water-sharing system. In case of water scarcity, the implementation of a quota system is necessary due to the high marginal value of water during the irrigation period. The collective regulation has to be based on a sound *quota + price* contracts between the service provider and the clients, which will leave each farmer free to manage his irrigation efficiently according to his own water value function. However, in a closing basin, the new resources must also be developed to ensure the governability of the system. The success of controlled water management, as developed in the Neste system, has been founded on joint management of demand and resources, with the construction of some new reservoirs during the last 10 years. The shift to an organizational set-up like the one described here can succeed, if governance, after the period of 'decision without consultation', avoids the current tendency towards 'consultation without decision'.

References

Balas, B. and Deumier, J.M. (1993) De l'entrée parcellaire à une stratégie globale dans la gestion de l'irrigation. *Aménagement et Nature* No. 3.

FAO (1996) NGO consultation on the world food summit. FAO, Rome.

Kosciusko-Morizet, N., Lamotte, H. and Richard, V. (1998) Que peut-on attendre de la mise en place de quotas individuels échangeables de prélèvement sur la ressource en eau en France? L'exemple de l'agriculture irriguée. Colloque SFER-CEMAGREF-ORSTOM, Montpellier.

Lesbats, R., Tardieu, H. and Heritier, M. (1996) Associations d'irrigants dans le sud ouest de la France, et leur relations avec la CACG. Question 46–1, paper R-107, ICID Congress, Cairo.

Malano, H. and Van Hofwegen, P. (1999) *Management of Irrigation and Drainage Systems. A Service Approach.* A.A. Balkema, Rotterdam/Brookfield.

Molden D.J., Sakthivadival, R., Perry, C.J., de Fraiture, C. and Kloezen, W.H. (1998) *Indicators for Comparing Performance of Irrigated Agricultural Systems.* Research Report 20. International Water Management Institute, Colombo, Sri Lanka.

Plantey, J., Tardieu, H., Mesny, M., Nico, J.P., Rieu, Th. and Verdier, J. (1996) Gestion de l'eau pour l'agriculture en France: durabilité socio-économique et implication des usagers. www.afeid.montpellier.cemagref.fr

Tardieu, H. (1991) Gestion de l'eau en Gascogne: la CACG, instrument technique et lieu de concertation. POUR No. 127–128. L'Harmattan, Paris.

Tardieu, H. (1999) La valeur de l'eau en agriculture irriguée: une information économique nécessaire pour mieux réguler la gestion de l'eau et des productions agricoles dans un marché ouvert. Symposium, ICID Congress, Granada.

Tardieu, H. and Plantey, J. (1999) Balanced and sustainable water management: the unique experience of the regional development agencies in Southern France. *ICID Journal*, no. 1.

Tardieu, H. and Préfol, B. (2002) Full cost or 'sustainability cost' pricing in irrigated agriculture. Charging for water can be effective, but is it sufficient? *Irrigation and Drainage* 51, 97–107.

7 Basin Management in a Mature Closed Basin: the Case of California's Central Valley

Mark Svendsen

7.1 Introduction

7.1.1 Basin management

River basins are managed at two different levels. At the higher level, the basin level, overall policies and plans are set, resources are allocated and regulations written and enforced. At the use level, regulated water deliveries are made to users of water, which may be irrigators, urban residents, industries, wetlands or natural river reaches. This paper focuses on the first level of management, the basin level, and examines the way in which basin level management functions are performed in the large interior Central Valley of California. The Central Valley comprises what T. Bandaragoda (unpublished note, 1999, International Water Management Institute (IWMI), Colombo, Sri Lanka) terms an *advanced* river basin, one which is already well developed in terms of physical infrastructure and effective institutions for integrated water resource management.

Issues of particular interest here are the interplay of political forces which support alternative water uses, the currently changing priorities accorded to alternative water uses, and processes and institutions whereby allocational and regulatory activities at the basin level are directed and coordinated. One central issue of global significance is the extent to which one apex organization must be in control of the highest level of decision making in a basin. Berkoff (1997), for example, has asserted that 'if water is to be managed holistically, all aspects must be coordinated by one . . . agency'. The present study suggests that this assertion does not apply universally and raises questions about the conditions under which different models of basin-level management would be most effective.

7.1.2 California's Central Valley

California's Central Valley is home to millions and is one of the premiere agricultural regions in the USA, containing six of the top ten agricultural counties in the country. California itself has 33 million residents and is the most populous state in the nation. An overwhelming 97% of the population live in urban areas.

The state as a whole has abundant renewable water resources, which, in addition to meeting environmental, urban and agricultural needs, generate 42% of the utility-produced electricity in the state. Irrigated agriculture generates 81% of California's total agricultural revenue on 30% of the state's farmland. Agriculture also provides 14.4% of the state's employment, though only 2.1% of that is engaged in direct production activities. The remaining 12.3%

works in input production, marketing and processing, and wholesale and retail sales. The state is also blessed with a magnificent and varied natural environment – the Pacific coast, the Sierra Nevada mountains, broad inland valleys, wetlands and the southern deserts. All of these features – the environment, urban concentrations, power generation and agriculture – require water for their sustenance and operation.

Several features of California's situation make it especially valuable as a case for study.

- First, California comprises a sophisticated economic environment in which water is used for a wide variety of purposes and is treated more as a commodity than as a common pool resource.
- Second, intense competition over water has emerged in what Seckler (1996) would call a *closed water system* – one in which there is little new water left to develop. This competition includes agricultural, municipal and industrial (M&I), and environmental interests and is driving rapid change in the institutions that allocate, regulate, convey and use water.
- Third, the responses to changing public priorities have been characterized by pragmatic problem-solving behaviour. This has made California a virtual laboratory for innovative solutions to problems of water reallocation and management, environmental quality, efficient water use and water quality management.

7.2. Basin Hydrology[1]

7.2.1 Supply

California possesses abundant water resources, receiving nearly 250 billion (10^9)

m^3 of precipitation annually in average years. Of this amount, about 65% is used by trees and other natural vegetation. An additional 10% flows to the Pacific Ocean or other salt sinks unchecked and unallocated. The remaining runoff is available as a renewable water supply for urban, agricultural and environmental uses.[2]

Developed surface water resources in the state total about 80 billion m^3, of which nearly half are set aside as required environmental flows.[3] About 12% of the total has been developed under Federal Government projects, 5% by the State of California and 17% by local government entities. An additional 8% comprises water imported from the Colorado River Basin under a multi-state water sharing agreement using facilities also constructed by the Federal Government.

In addition to surface water sources, an additional 15 billion m^3 is available as renewable groundwater. Present withdrawal rates are higher than this, resulting in an overdraft of about 1.8 billion m^3 annually, some 12% of the renewable total. Furthermore, the rate of overdraft is increasing and was 10% greater in 1995 than it was in 1990. To some extent, this overdrafting is a consequence of 1992 federal legislation reallocating water away from irrigators to environmental uses. This has led to supply deficiencies of up to 50% for some Central Valley irrigators and caused them to turn to lightly regulated groundwater as a replacement supply.

Most of California's precipitation falls as snow in the mountains of northern California and in the Sierra Nevada range, which comprises the high backbone of the state running from north to south along its eastern flank (Fig. 7.1). A second range of much smaller hills, the Coastal Range, fronts the narrow coastal plane in the West, creating a broad alluvial valley between the two ranges. This Central Valley is an

---

[1]   Data for this section are drawn largely from DWR (1998).
[2]   A portion of the water specifically designated for in-stream environmental use also flows to the Pacific Ocean.
[3]   The total of developed surface and groundwater is greater than the 25% of precipitation designated as available runoff because of reuse.

Fig. 7.1. California's Central Valley.

area of rich soils and favourable growing conditions for a wide variety of crops and is the heart of California agriculture. In it, more than 200 types of crops are grown and from it comes 45% of the nation's fruits and vegetables. Two major river systems drain the Central Valley and some 158,000 km² of watershed, the Sacramento River in the north, and the San Joaquin River in the south. The two rivers meet in the Sacramento–San Joaquin Delta (the Delta), just inland of San Francisco Bay, from where they flow into the Bay and out to the Pacific Ocean.

The fact that two-thirds of California's water is in the north, while the bulk of agricultural land and the largest population centres are in the south, has led to two massive engineering projects designed to transport water from north to south. These are the federal Central Valley Project (CVP) and the State Water Project (SWP).

The CVP was constructed in the 1940s by the Bureau of Reclamation, the federal irrigation development agency. Construction was begun in 1935 as a part of a massive depression-era public works programme. The project is anchored by Shasta dam in the Cascade Mountains in northern California, which stores water for use in the south. Water from Shasta and several smaller dams is routed down the Sacramento River to the Delta, which it crosses in a network of natural and artificial channels. Some of the water is used to irrigate land along the Sacramento River to the north, but most crosses the Delta to be lifted 60 m into the Delta–Mendota Canal (DMC). The DMC supplies 32 irrigation districts in the San Joaquin Valley with water.

The second project, the SWP, was developed in the 1960s by the State of California. Its backbone, the California Aqueduct, parallels the DMC south from the Delta before continuing on to southern California. Its primary purpose is to convey M&I water to desert cities in the south (70%), principally the greater Los Angeles area, though it does supply irrigation water (30%) as well. Together these two projects deliver about 7.3 billion m³ of water annually to the south.

7.2.2 Demand

7.2.2.1 Current patterns

Overall demand for developed sources of water is dominated by environmental reservations (46.5%) and by irrigated agriculture (42.5%) (Table 7.1). Municipal demand currently makes up 11.0% of the total.

7.2.2.2 Changing patterns of demand

Projections for 2020 (DWR, 1998) anticipate only a very modest expansion in available supply (1%), but with important shifts in the composition of use. While environmental uses of water are expected to remain

Table 7.1. Average water year water use, 1995 and 2020.

	1995		2020	
	Volume [10⁹ m³]	Share	Volume [10⁹ m³]	Share
Urban	10.8	11.0%	14.8	14.9%
Agricultural	41.7	42.5%	38.9	39.1%
Environmental	45.6	46.5%	45.6	45.9%
Total	98.0	100.0%	99.3	100.0%

constant, urban demand will expand by 37% and agricultural water use will shrink by nearly 7% to accommodate this growth. Additional developed supplies will be devoted entirely to urban use.

The federal Endangered Species Act, passed in 1973, established the legal framework for protecting species of plants and animals listed as *threatened* or *endangered* and the allocation of water for their preservation where necessary. The listing of winter run Sacramento River salmon as endangered under this act in the early 1990s was the first important application of the law in California, which had a significant impact on water allocation in the Sacramento–San Joaquin Valley. A far more sweeping change was wrought by the Central Valley Improvement Act, passed by Congress in 1992. This act reallocated a portion of the water the federal government had contracted to deliver to irrigation districts to the ecosystems of the Sacramento–San Joaquin Delta. This reallocation has resulted in significant shortfalls in supplies to many of the irrigation districts in the San Joaquin Valley.

7.2.2.3 Urban use

Driving the growth in urban water use is projected growth in the California population of nearly 50% between 1995 and 2020, as a result of continuing in-migration from others regions of the country and from abroad. The demand for water caused by this growth completely overshadows modest potential reductions in per capita water use of about 6% if household level best management practices are fully implemented.[5]

7.2.2.4 Agricultural use

California has more than 3.6 million ha of agricultural land under irrigation, 80% of it in the Central Valley. Projections for 2020 indicate a modest reduction in the total irrigated area of about 130,000 ha (3.6%) resulting mainly from urban encroachment, land retirement due to drainage problems, and more competitive economic markets for agricultural products.[6] In addition, changes in cropping patterns and irrigation technology and practices will yield small reductions in the rate of per hectare water use (an estimated 2.4% of 1995 use levels).

7.2.2.5 Environmental use

Environmental water use comprises several categories of flows that have been set aside for environmental purposes. These are:

- Dedicated flows in designated 'wild and scenic' rivers (64%);
- Instream flow requirements in other rivers established by water right permits, court actions, agreements or other regulatory actions (17%);
- Required Sacramento–San Joaquin Delta outflows (15%);
- Wetlands freshwater requirements (4%).

Note that while there are other environmental uses of water, the above uses are distinguished by being managed and quantifiable. Most of this environmental water allocation is brought about by legislative and regulatory processes rather than through the water right permitting process, which authorizes agricultural and municipal uses.

7.2.3 Summary

California is well endowed with renewable water resources. Of the 250 billion m^3 received as precipitation annually, about one-quarter is available for various allocated uses. About half of this allocated water is set aside for instream environmental uses. The remainder (just over 50 billion m^3) is available for withdrawal for agricultural and urban uses. Groundwater, though abundant, is currently overdrafted by about 12% of the renewable total and exploitation continues to expand.

Two major plumbing projects, one federal and the other state, transfer water from the wet north to the arid south of the state. Water moving through both of these systems must transit the Sacramento–San Joaquin Delta 'in the open', where it mixes with water in the Delta and contributes to it. The Delta is also important environmentally, and it serves as the nexus of the debate over the future of California's water.

According to the most recent version of the California Water Plan, urban demand is expected to grow by 37% over the next quarter-century, while agricultural water use shrinks by 7% and environmental use holds constant. Additional allocations to environmental uses are being promoted, however, and if they are adopted, additional reallocation of agricultural water will be the likely outcome.

7.3 Legal, Policy and Institutional Environment

7.3.1 Water rights

Water in California, as in the USA in general, is regarded as a good belonging to all and held in trust by the State.[7] Management of water, and allocation of rights to use

---

[5] If explicit conservation practices are not implemented, per capita urban demand will increase by about 6%.

[6] There is potential for much more significant reductions if major proposed conversions of agricultural land to wildlife habitat are implemented.

[7] This is the Public Trust Doctrine, derived from Roman Law.

water, are the responsibility of the individual states. Rights to use water in California comprise a complicated mixture of types, priorities and levels of security. Ground and surface water rights are treated separately, and surface water rights, which are the most important, include both *riparian* and *appropriative* rights. Underlying and articulating the various elements of the allocation scheme are a number of state and federal laws and numerous court cases, each of which establishes precedents upon which subsequent cases build.

Riparian rights to surface streams are available, under common law, to the owners of property abutting streams. Water abstracted under a riparian right cannot be applied to plots of land that do not abut the stream, and cannot be transferred to other uses removed from the riparian land. They comprise about 14% of rights to non-imported surface water in California.

Appropriative rights to surface water are more flexible and comprise the remaining 86% of non-imported surface water rights. Appropriative rights are granted through a permitting process managed by the State of California. Appropriative rights can be for use at points removed from the stream of origin and are subject to transfer and change of purpose. Maintenance of an appropriative water right requires continuous beneficial use, and the courts have held that appropriative rights can be lost after 5 years of non-use. Riparian rights are neither created by use nor lost by non-use.

Groundwater use is only lightly regulated. There is no permitting process for groundwater exploitation, which is available, in the first instance, to owners of overlying land for reasonable beneficial use on those lands. Groundwater users establish rights simply by use. Rights are correlative with the rights of other owners, meaning that if the water supply is insufficient, the supply must be equitably apportioned. Subject to future requirements on overlying lands, 'surplus' groundwater may be appropriated for use on non-overlying lands. Again, no permit is required.

This very vague and permissive specification of rights to groundwater has two important implications. First, as pressure on nearly fully allocated surface water sources continues to build, users turn to groundwater to make up deficits, leading to a serious and growing problem of overdrafting in many portions of the state. Second, groundwater is a magnet for litigation as water users joust over such terms as 'surplus', 'sufficient', 'reasonable', 'equitable' and 'beneficial'. Development of a suitable institutional framework for managing groundwater in the state is urgently needed but proceeding slowly.

7.3.2 Actors

There are seven important groups of actors involved in basin-level water management in California, in addition to the general public. These are the managers, the service providers, the users, the regulators, advocacy groups, elected officials and the courts. Some groups, such as regulators, service providers, the courts and elected officials, consist of both federal and state-level actors, while others are purely local. The main actors in each category are discussed briefly below.

7.3.2.1 Managers

The most important managing organizations are two California state organizations – the Department of Water Resources (DWR) and the State Water Resources Control Board (SWRCB). The DWR replaced the Office of the State Engineer in 1956, assuming responsibility for planning and guiding development of the state's water resources. Over the past 45 years, it has grown from 450 staff to a level ten times that in 1967 before dropping back, and now employs about 2000. DWR operates on an annual budget of about $1 billion and is a division of the state public administration under a director who is accountable to the state governor. Its responsibilities are primarily technical and operational, but do include some regulatory functions. Major responsibilities include the following.

- Preparing and updating the California Water Plan every 5 years;
- Operating and maintaining the SWP;
- Protecting and restoring the Sacramento–San Joaquin Delta;
- Dam regulation and flood protection;
- Public education;
- Providing technical assistance to local communities.

The SWRCB performs functions that are managerial, regulatory, and quasi-judicial in nature. It thus occupies a special niche in the overall set-up. Among its important responsibilities are the following:

- Allocating rights to appropriate (use) surface water;
- Adjudicating disputes over rights to water bodies, such as the Sacramento–San Joaquin Delta;
- Establishing water quality standards;
- Guiding and overseeing the nine Regional Water Quality Control Boards.

Board appointments are made by the governor and the makeup of the Board is as described in Box 7.1.

Regional Water Boards under the SWRCB do not allocate water rights but manage and regulate water quality through the following kinds of actions:

- Writing waste discharge permits;
- Implementing contamination clean up operations;
- Monitoring quality and use of regional ground and surface waters;
- Inspecting dischargers and enforcing state and federal water quality laws.

Regional Boards consist of five members who are also appointed by the governor.

7.3.2.2 Service providers

At the basin level, the most important water service providers are the United States Bureau of Reclamation (USBR or 'the Bureau') and DWR. The USBR is an agency of the Federal Government housed in the Department of the Interior. The Bureau constructed most of the federally financed water conveyance and control facilities in the state, including the pivotal CVP, and operates the storage and delivery facilities it has constructed. However, while it retains operating responsibility for the upstream portions of the CVP, it has recently transferred operating responsibilities for the portions of the system lying south of the Delta to an organization established and controlled by San Joaquin Valley water users, the San Luis–Delta Mendota Water Authority (SLDMWA). Users have proposed that they assume responsibility for the upstream portions as well, but action on that step is more controversial.

The other major water storage and conveyance project in the state, the SWP, is operated by DWR, which constructed the facilities using state resources.

7.3.2.3 Users

Principal water users are the various districts that purchase water and deliver it to the members or residents in the district. Districts are generally organized to supply irrigation water to farmers or municipal water to urban residents. Districts are incorporated as non-profit entities under state law and are self-governing. The largest share of managed surface water is delivered to agricultural users, most of whom are in the Central Valley. Other users include the state Department of Fish and Game, conservation districts, hydropower facility operators, and DWR and the USBR for flood control operations. Freshwater navigation, though significant in the past, is of minor importance today.

7.3.2.4 Regulators

Water-related regulation centres around provisions of federal and state laws protecting endangered species and maintaining drinking water quality. The federal Endangered Species Act of 1973 is the most important of these and is enforced by the national Environmental Protection Agency (EPA). Of the endangered species affecting water use in California, the most critical are the listed runs of salmon. Technical regulations and certifications relating to salmon

---

Box 7.1.   Evolution of a water control agency.

1940s	Serious water-quality problems emerged in California, including outbreaks of water-borne diseases and degradation of fishing and recreational waters. In 1949, a fact-finding committee highlighted cumbersome and unreasonable laws and administrative procedures, multiple jurisdictions, limited and conflicting interests, and overlapping authorities as roots of the evident problems. The committee concluded that the state's limited water resources could only be extended through planning to maintain water quality while at the same time allowing maximum economic use and reuse. It recommended a central focus point at the state level to coordinate water pollution control activities.
1949	Legislation created a **State Water Pollution Control Board** consisting of nine gubernatorial appointees representing specific interests and four *ex officio* state officials. Its duties included formulating statewide policy for pollution control and coordinating the actions of various state agencies and political subdivisions of the state in controlling water pollution. The same legislation created nine **Regional Water Pollution Control Boards** in major watersheds. These regional boards had responsibility for administration, investigation and enforcement of the state's pollution abatement program. Five gubernatorial appointees, representing water supply, irrigated agriculture, industry, and municipal and county government in the region, served on each regional board.
1959	The 1949 law was revised and broadened on the basis of 10 years of experience. State *ex officio* members were removed from the board, increasing its separation from the state administrative machinery.
1963	The state board was renamed the **State Water Quality Control Board** and given the broader mandate of water quality control, replacing the more limited earlier focus on sewage and industrial waste control.
1967	A proposal to consolidate water-related functions, including water quality control functions, within the Department of Water Resources (DWR) was rejected on the grounds that this would create conflicts of interest internal to DWR. Instead quantity and quality management functions were consolidated external to DWR by merging the State Water Quality Control Board and the State Water Rights Board into the **State Water Resources Control Board**. The 'State Water Board' consists of five full-time members mandated to protect water quality and to determine rights to surface water use. Members are appointed by the governor and fill specialized roles on the Board, i.e. an attorney versed in water law, two civil engineers with expertise in water rights and water supply, a water quality member, and a public member.
1969	A new Water Quality Control Act was passed which retained the basic structure of state and regional boards but provided a new regulatory framework for waste discharges to both surface and ground waters. This act served as a model for the federal Clean Water Act, passed 3 years later.

Source: http://www.dwr.water.ca.gov

---

are made by the National Marine Fisheries Service (NMFS), whereas criteria for other animal species are set and supervised by the federal Fish and Wildlife Service.

State environmental regulators also list endangered species, and this list includes some that are not on the federal list. The state Fish and Game Department supervises enforcement of water quality and quantity requirements relating to state-listed species.

The SWRCB and its subtended regional boards bear overall responsibility for surface and groundwater quality in the state. The Federal Clean Water Act and the California Water Quality Control Act, both aimed at pollution control, are enforced by these boards.

7.3.2.5 Advocates

One of the most dramatic recent changes in the cast of characters in the water drama in California, and in the USA, is the emergence of environmental advocacy groups as potent political actors. Most groups are membership based and supported and may draw on grants from charitable foundations. Some focus on a single issue – a resource or species – while others have a broader range of interests. The group Friends of the River is an example of a resource-focused group which is largely concerned with restoring free-flowing rivers in California, while the Sierra Club, based in California but national in scope, is an example of a group with a wide range of conservation interests beyond water. There are about 20 environmental groups in California interested in water issues. These groups are linked through an Environmental Water Caucus, which meets every couple of weeks. Accompanying expanded federal and state environmental regulation over the past 25 years has been greatly strengthened requirements for transparency in regulatory processes.

7.3.2.6 Elected officials

Legislators at both the federal and state levels write the laws providing the framework for water resource management in the state. Although establishing systems for allocating water resources is the purview of the state legislature, the federal government exerts a powerful influence on water allocation by applying the terms of the federal Endangered Species Act. This act constrains water-related construction projects in various ways, and can require increased in-stream allocations of water for fish species classed as *threatened* or *endangered*. The governor is a particularly important figure in the state water resource management picture, controlling appointments to the State Water Board and the regional Water Quality Boards and as the head of the state administrative apparatus, which includes the important DWR. The US Congress also influences allocation through its ability to mandate changes in water permits, which are held on behalf of the US Government by the US Bureau of Reclamation. Water quality is regulated by both federal and state statutes.

7.3.2.7 Courts

Both state and federal courts hear cases relating to water. Where the US Government is a party to the litigation, a federal court must be the venue, as state courts cannot have jurisdiction over the Federal Government. Almost all of the cases heard are civil cases rather than criminal cases, involving disputes between parties rather than a violation of state or federal law.

A prominent federal judge in the Central Valley estimates that about 20% of his caseload consisted of water cases, and that the volume of water-related cases had increased considerably over the past 9 years. Cases have also increased in complexity. The introduction of the Federal ESA and the listing of a number of fish species in important California rivers has played a major role in this increased complexity. The integration of Public Trust Doctrine into California water law[8] has also made decisions more complicated. Throughout this period of change, the non-governmental (NGO) sector has increasingly become a 'third presence' in nearly every significant civil case, seeking to include the environmentalist viewpoint into the deliberation.

Major drawbacks to the heavy reliance on the court system for dispute resolution are the often drawn-out nature of proceedings, their expense and the difficulty of reaching sound decisions through adversarial proceedings. A federal judge interviewed cited approvingly an old adage, 'hard cases make bad law'. Increasingly attention is shifting to various modes of alternative dispute resolution.

7.3.3 Essential functions

Burton (1999) has identified 11 essential functions of basin management. A somewhat modified listing of these functions is shown in Table 7.2, crossed with the key actors identified in the previous section.

---

[8] Accomplished by a decision of the state Supreme Court in 1983 in a suit filed by the National Audubon Society.

Table 7.2. Essential basin management functions and key actors.

Key actors	Surface water									Groundwater							Waste water					Ag. returns			
	Plan (basin-level)	Allocate water	Construct facilities	Distribute water	Maintain facilities	Monitor quality	Ensure quality	Protect against flooding	Protect ecology	Plan (basin-level)	Allocate water	Construct facilities	Withdraw/distribute	Maintain facilities	Monitor quality	Ensure quality	Construct facilities	Operate facilities	Maintain facilities	Monitor quality	Enforce quality	Construct facilities	Maintain facilities	Monitor quality	Enforce quality
Dept of Water Resources	●		●	●				●		●															
WRCB/RWQCBs		●				●	●		●						●	●				●	●			●	●
USBR	○		●	●	●					○												●			
SL-DM Water Authority			○	●	●	○																			
Irrigation Districts	○		●	●	●					○		●	●	●								●	●	●	
Municipal Water Districts	○			●	●	●				○		●	●	●			●	●	●			●	●	●	
Industries	○				●					○		●	●	●			●	●	●						
US EPA/NMFS/FWS									●																
CA EPA/FGD							●		●							●									
Advocacy groups (NGOs)	○					○	●		●	○					○	●									
Courts		●							●												●				●
CALFED	●								●																
US Congress	○	○																							

●, Indicates activity; ○, indicates limited activity.

WRCB, Water Resources Control Board; RWQCB, Regional Water Quality Control Board; USBR, United States Bureau of Reclamation; US EPA, US Environmental Protection Agency; NMFS, National Marine Fisheries Service; FWS, US Fish and Wildlife Service; CA EPA, California Environmental Protection Agency; F&GD, California Fish and Game Department; NGO, (environmental) non-governmental organizations; CALFED, California/Federal Bay-Delta Program.

These functions are replicated, as appropriate, across four broad categories – surface water, groundwater, wastewater disposal and agricultural return flows. Cells are marked to indicate an actor active in a particular functional area. Information is drawn from interviews, printed materials and Internet postings. A number of interesting points emerge from an examination of Table 7.2.

- A comprehensive planning function rests with the state DWR. This responsibility covers surface and groundwater and both quantity and quality. Although technical analyses and modelling are done by DWR, extensive interaction with a variety of stakeholders in the planning process makes planning a widely shared activity. The primary planning document is the State Water Plan, which is updated every 5 years in a process led by DWR.
- Surface water allocation and water quality assurance are assigned to a single state agency, which is independent of the other state agencies engaged in planning or system operations. The WRCB is autonomous, though it is political to the extent that the members are appointed by the governor of the state. The US congress assumed a certain amount of *de facto* allocational authority in passing a 1992 law, which directed the USBR to reallocate water from agricultural users with whom it held contracts to environmental uses. Federal and state courts also play important roles in the allocational process by resolving disputes over allocation.
- Enforcement of water quality standards rests with nine regional boards with strong local ties but under the overall guidance of the state-level WRCB. The courts also play significant roles in interpreting disputes related to water quality.
- Retail water delivery services are, for the most part, in the hands of user-controlled districts. Such irrigation and municipal water supply districts are financially autonomous and self-regulating. They usually obtain water from wholesale suppliers through legally enforceable contracts.
- Groundwater is the most lightly planned and regulated segment of the state's water resources. There is little control over abstractions, and the state is in a serious overdraft situation.
- Advocacy groups (environmental NGOs) make up an important third presence in most important disputes involving water. This is a relatively recent development, but has profoundly changed the way in which decisions are made, and modified their outcomes. These groups also play important roles in joint consensual processes, such as CALFED and the American River Water Forum, which are being used increasingly to develop mutually acceptable plans and agreements over contentious water-related issues.
- There is a certain internal conflict within the DWR with respect to its multiple roles as wholesale supplier of water, water resource planner and regulator. Transparency of process appears to keep these potential conflicts in check. Although not included in the table, there is a significant conflict of interests internal to the US Army Corps of Engineers, which is charged with wetland permitting and protection, but is, at the same time, a construction and operating agency with close ties to the congressional appropriations process.

7.3.4 Coordinating processes

Managing an important publicly held natural resource will always involve multiple actors, differing interests and perspectives, and relational dynamics. This is true even in situations where a single agency is responsible for all aspects of basin water management, as there will be winners and losers among users of basin water resources and factions within the managing agency having differing perspectives and interests.

In California, where there are many discrete actors in the water resource allocation and management picture, coordination and decision making have long been important and contentious functions. Traditionally, the courts, both federal and state, have provided a critical dispute resolution function. As the Sacramento–San Joaquin basin has closed and water becomes relatively scarcer, disputes have become more frequent and the number of interested parties has grown, making proceedings more complex. There is presently growing interest in various forms of alternative dispute resolution, including the use of mediation, arbitration and special masters.

There is also growing reliance on processes of shared consensual decision making to replace the more typical two-stage process of a technical decision made by a government agency, followed by extensive and lengthy litigation initiated by unsatisfied parties. The most prominent example is the ongoing CALFED process, which tackles some of the most contentious water-related problem in the state, as shown in Box 7.2.

CALFED is a consortium of federal and state government agencies with management and regulatory responsibilities in the Bay–Delta system. It was formed in 1994 with the mission of developing a long-term comprehensive plan to restore ecological health and improve water management for beneficial uses of the Bay–Delta system, the heart of the Central Valley hydraulic system. CALFED spent its first 2 years identifying and defining problems and a further 4 years assessing the environmental implications of various actions that might be taken. It is about to begin an implementation phase that could last 30 years and cost $10 billion.

What sets CALFED apart from other programmes is the fact that problems and solutions are being discussed from the outset in an open forum with participation that spans the entire range of water-related

interests, and that it is proposing an entire basket of measures which will address the four problem areas in an integrated, complementary, sustainable way. Fundamental principles guiding the process are shown in Box 7.3. What is striking is the commitment of all participating parties to make the CALFED approach work. This commitment arises in part from the fear that if the process fails, years of litigation will follow in a far more adversarial and expensive process of dispute resolution.

7.3.5 Enabling conditions

The essential functions and actors' roles depicted in Table 7.2 provide a static view of responsibilities. Additional attributes of well-functioning basin governance[9] systems relate to their dynamics. We term these attributes, which provide the context for functional performance, *enabling conditions*.

Box 7.2.   Bay–Delta problem areas.

- Ecosystem restoration
- Water quality assurance
- Levee system improvement
- Water supply reliability

Box 7.3.   CALFED solution principles.

Affordable – solution can be implemented and maintained with the foreseeable resources of the CALFED stakeholders
Equitable – solution will focus on resolving problems in all problem areas
Implementable – solution has broad public acceptance, legal feasibility and will be timely and relatively simple compared with alternatives
Durable – solution will have political and economic staying power
Reduce conflicts – solution will reduce major conflicts among beneficial users of water

---

[9]   The term *governance* is used in a somewhat different sense here than in Burton's list of essential attributes, of which it is one. Here the term refers to the rules providing the context for multi-actor basin management and the processes and activities engaged in by those actors operating within this set of rules.

Enabling conditions are features of the institutional environment at the basin level that must be present, in some measure, to achieve good governance and management of the basin. These attributes are not specific to any one actor, but apply to all actors and their interactions and comprise necessary (but not sufficient) normative conditions for success. Basic enabling conditions are shown in Box 1.2 in Chapter 1. While a full analysis of these factors is well beyond the scope of this chapter, a brief sketch of each, in the context of California, is given to illustrate the concepts and indicate broad strengths and weaknesses.

7.3.5.1 Political attributes

Representation is generally well developed, with groups having similar interests allied into various associations. These associations are supported by their members who provide funds for representation and litigation. Environmental concerns are represented by NGOs, which have grown over the past 25 years in number, resources and influence. Supported by protections provided by federal and state endangered species laws, they now enjoy power commensurate with the other major players.

7.3.5.2 Informational attributes

The availability of information and transparency of decision-making processes in the USA, and in California, has also expanded over the past quarter-century. These changes have been driven by requirements in environmental protection laws, by the existence of the World Wide Web, and by growing public demand for information and openness. It is now a rare decision-making process that is not characterized by ready availability of technical information, public hearings and extensive opportunities for public comment.

7.3.5.3 Legal authority

The system of surface water rights, though complex, is relatively well specified in law and through cumulative court decisions.

Rights to groundwater are not well specified and comprise an area where a stronger and more appropriate legal basis is urgently required.

There is a sound legal framework underlying user-based districts, which provide such services as irrigation, domestic water supply, groundwater management and wetland conservation. Districts are self-financing and self-governing and generally work effectively.

7.3.5.4 Resources

Though participants always feel that financial resources are inadequate, both financing and human resources within the basin management system appear generally adequate. There is a well-developed physical infrastructure for transferring water around the state, and from neighbouring basins, and a steady stream of additions and improvements to it. Environmental restrictions and concerns, however, make infrastructural design a far more demanding process than it previously was, and have completely stymied some proposed projects, such as the *peripheral canal* around the Delta. New institutional forms (along with a legal basis for them) will likely be required in the future to legitimize and implement consensual agreements reached by *ad hoc* bodies such as CALFED, but the need for these is still evolving.

7.4 Salient Characteristics of California Basin Management

A number of important features characterize basin water management in California. These are summarized below:

- Multiple sources of authority and power – no single public agency manages water resources in California's river basins. Instead, decisions are made and enforced by a number of state and federal agencies. Integration is provided by the State Water Plan, various regional plans and processes such as

CALFED, the centralized system of surface water rights and the court system.

- Dynamic interplay of competing interests – an even broader group of actors participate in and influence decision making. These actors are from both public and private sectors. They debate in a variety of fora to assert their points of view. These include public hearings, the media and the courts. Extensive lobbying of public officials also takes place behind the scenes. Decisions emerge from this interplay.
- Adequate representation of all interested parties – major parties in the water debate are well represented and financed. These include municipal water districts, agricultural water districts, public water supply agencies, state and federal environmental regulators, and environmental NGOs.
- Heavy reliance on legally enforceable contracts and agreements – many of the water-related decisions made take the form of contracts or agreements between two or more parties, rather than administrative decree. This requires confidence on all sides in the enforceability of the agreements.
- Separation of operating and regulatory functions – regulatory functions are generally handled by organizations which are independent of federal, state and user-controlled operating agencies.
- Adequate databases on hydrologic processes and capacity to research new issues – extensive measurement and data collection programmes have created a large database of information on California water resources and their uses and impacts. Equally importantly, a strong technical capacity exists in the private sector to conduct additional assessments, on a consulting basis, as needs arise.
- Open access to information and generally transparent decision-making

processes – information on water flows, water quality, wastewater quality, water rights and so on is available to the public and is generally accessible through the World Wide Web, and in publications and public records. Decision-making processes are generally conducted in the open and include public hearings.[10] Moreover, decisions reached are subject to challenge in court and decisions over controversial issues often are so challenged.

- Self-financing autonomous districts as retail service providers – retail water service delivery is typically handled by irrigation or water districts, which are user-controlled, self-financing, non-profit quasi-municipal entities incorporated under state law. This vastly simplifies the service delivery problem by reducing the number of major 'users' to several hundred from tens of thousands.
- Important role of an impartial court system in resolving disputes – federal and state courts are regularly called upon to settle disputes brought to them as civil suits. Without this service, the water resource management system in the state would be unworkable.
- Well-defined system of water rights (except groundwater) – there is a clear system of allocating and protecting rights to surface water, which provides reasonable security to users. Protection of groundwater is presently more problematic.

References

Berkoff, J. (1997) *Water Resources Functional Analysis*. Report prepared for the Sri Lanka Water Resources Council, Colombo, Sri Lanka.

Burton, M. (1999) Note on proposed framework and activities. Prepared for the

---

[10] As is always the case, real compromises are often hammered out in private by a smaller group of participants. Nevertheless, the compromises reached must be capable of standing up to public and interest group scrutiny when they are announced.

IWMI/DSI/CEVMER Research Programme on Institutional Support Systems for Sustainable Management of Irrigation in Water-Short Basins, Izmir.

Department of Water Resources (1998) California Water Plan Update Bulletin 160–98.

Department of Water Resources, Sacramento, California.

Seckler, D. (1996) The New Era of Water Resources Management: From 'Dry' to 'Wet' Water Savings. IWMI Research Report 1. IWMI, Colombo, Sri Lanka.

8 River Basin Closure and Institutional Change in Mexico's Lerma–Chapala Basin

Philippus Wester, Christopher A. Scott and Martin Burton

8.1 Introduction

The Lerma–Chapala basin in central Mexico is a telling example of the institutional and political challenges that river basin closure poses, especially for locally managed irrigation. Although rainfall in the past 10 years has been slightly above average, total water depletion in this basin exceeds supply by 9% on average. Groundwater is being mined, with sustained declines in aquifer levels of 1.00–2.58 m/year (Scott and Garcés-Restrepo, 2001), while surface water depletion exceeds supply in all but the wettest years. As a result, Lake Chapala, the receiving water body of the basin, is drying up. In early 2001, the volume of water stored in the Lake was around 20% of capacity, the second lowest level recorded since systematic data collection began in 1934. This lake is the largest in Mexico, giving it a high symbolic value, and it generates significant tourism revenue.

Approximately 68% of the annual renewable water in the basin is used to irrigate around 700,000 ha, all of which is locally managed. Since the basin's water is fully committed, there is no scope for irrigation expansion, and the drilling of new wells and the construction of new dams has been prohibited. Moreover, water pollution is serious, with significant wastewater reuse for irrigation within the basin. Lastly, water is being transferred from agriculture to the urban and industrial sectors, without

due compensation to farmers. In sum, the sustainability of locally managed irrigation in the basin is under serious threat, carrying with it grave implications for social equity. Due to basin closure, the poorer segment of the farming community is losing its access to primary water sources as it is diverted to economically higher-valued uses. Consequently, poor farmers increasingly vie for derivative water such as wastewater and drainage effluent (cf. Buechler and Scott, 2000).

In response to the water crisis in the basin, several institutional changes have occurred in the basin since 1989, including the signing of a river basin coordination agreement (1989), the creation of a River Basin Council (1993) and the establishment of aquifer management councils (1995 onwards). Water reforms at the national level, such as the creation of a national water agency (1989), the decentralization of domestic water supply and sanitation to state and municipality levels (starting in 1983), the transfer of government irrigation districts to users (1989 to present), the creation of state water commissions (from 1991 onwards), and a new water law (1992), have also significantly altered institutional arrangements for water management in the basin (González-Villareal and Garduño, 1994). The water reforms in Mexico are driven by the need to deal with increasing water over-exploitation, and influenced by the vested interests of the hydraulic bureaucracy and the neo-liberal economic policies pursued by the Mexican government (Rap *et al.*, 2004).

This chapter assesses the institutional arrangements for water management in the Lerma–Chapala basin and how well they are dealing with basin closure. It does so by presenting an overview of the basin's water resources and uses, followed by a section exploring the legal, policy and institutional conditions that influence how the basin is governed and managed. This basin profile provides the backdrop for an analysis of the essential functions for river basin management, after which the key challenges facing the basin, namely surface and groundwater allocation mechanisms and the management of derivative water, are reviewed and conclusions for the future direction of institutional change in the basin are drawn.

8.2 The Lerma–Chapala Water Balance

The Lerma–Chapala basin covers some 54,300 km² and crosses five states: Querétaro

(5%), Guanajuato (44%), Michoacán (28%), Mexico (10%) and Jalisco (13%).[1] The basin is home to a dynamic agricultural sector and a growing industrial sector and accounts for 9% of Mexico's GNP. It is the source of water for around 15 million people (11 million in the basin and 2 million each in Guadalajara and Mexico City) and contains 13% of the irrigated area in the country. The average annual runoff in the basin from 1940 to 1995 was 5,757 million cubic metres (MCM), a little over 1% of Mexico's total runoff (CNA, 1999a).

The headwaters of the Lerma River rise in the east of the basin near the city of Toluca at an elevation of 2600 m a.s.l. (metres above sea level) to discharge into Lake Chapala in the west at an elevation of 1500 m a.s.l. The total length of the Lerma River is 750 km, and eight major tributaries discharge into it (Fig. 8.1). Lake Chapala, with a length of 77 km and a width of 23 km, is Mexico's largest natural lake, storing 8125 MCM and covering 111,000 ha when full. The shallow depth of the lake (average 7.2 m)[2] results in the loss of around

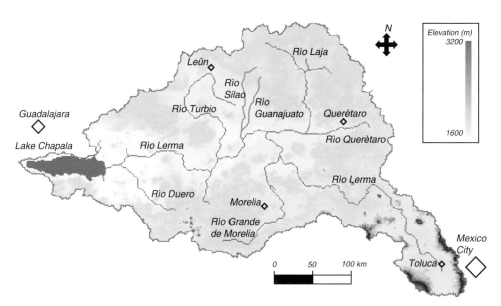

Fig. 8.1. Topography and stream network of the Lerma–Chapala basin.

---

[1]   Percentages indicate the area of the basin that falls in each state.
[2]   On a scale of 1 : 10,000 the dimensions of the lake are 7.7 m by 2.3 m and less than 1 mm deep.

1440 MCM (25% of the annual average runoff in the basin) of its storage to evaporation each year (de Anda *et al.*, 1998). At times of high water levels, Lake Chapala discharges into the Santiago River, which flows in a northwesterly direction and then drops to the Pacific Ocean after 524 km.

The climate in the basin is semiarid to subhumid, with 90% of the rain falling between May and October. Rainfall is highly variable, with an average over the 1945–1997 period of 712 mm/year, a minimum of 494 mm in 1999 and a maximum of 1022 mm in 1958 (CNA, 1999e). Average monthly temperatures vary from 14.6°C in January to 21.3°C in May; thus a range of crops can be grown throughout the year. The potential evapotranspiration mirrors the temperature variation, with a peak in April/May, and an annual total of some 1900 mm. In every month except July and August there is a net deficit between rainfall and potential evapotranspiration, indicating the importance of irrigation.

Forty aquifers have been identified in the basin (CNA/MW, 1999). The upper layer of these aquifers is generally 50–150 m thick and composed of alluvial and lacustrine materials, while the lower layers, several hundred metres in depth, are composed primarily of basaltic rocks and rhyolite tuff. The aquifers are recharged through rainfall infiltration, surface runoff and, importantly, deep percolation from surface irrigation. Various sources report different data on annual extraction and recharge rates, making it hard to portray with any precision the groundwater situation in the basin. What is clear is that 30 of the 40 aquifers are in deficit, with static water levels dropping at 2.1 m/year on average (Scott and Garcés-Restrepo, 2001). Recent data from the *Comisión Nacional del Agua* (CNA; National Water Commission) indicate that average annual recharge is 3980 MCM, while average annual extractions are placed at 4621 MCM, giving a deficit of 641 MCM, some 71% of the total water deficit in the basin (CNA, 1999a).

Table 8.1 presents current average consumptive water use for different sectors in the basin compared to average annual renewable water, and shows a deficit of 900 MCM. The percentage of available water that is developed and put to use in the basin is 109%, showing its degree of over-commitment. The out-of-basin transfers are to Guadalajara (surface water) and Mexico City (groundwater) for urban water supply. To portray basin closure in the Lerma–Chapala basin, it is instructive to analyse fluctuations in the water levels of Lake Chapala. Figure 8.2 shows these fluctuations from 1934 to 2000 and relates them to developments in the basin.

Starting in 1945, water levels in the lake declined sharply, from around 97 m on

Table 8.1. Water balance of the Lerma–Chapala Basin.

	Surface water		Groundwater		Total	
	MCM	%	MCM	%	MCM	%
Runoff/recharge	5,757	100	3,980	100	9,737	100
Total depletion	6,016	104	4,621	116	10,637	109
Irrigated agriculture	3,424	59	3,160	79	6,584	68
Urban	40	> 1	751	19	791	8
Out-of-basin transfer	237	4	323	8	560	6
Industry	39	> 1	239	6	278	3
Other	6	> 1	148	4	154	2
Total consumptive use	3,746	65	4,621	116	8,367	86
Evaporation from water bodies	2,270	39	–	–	2,270	23
Balance	−259	−4	−641	−16	−900	−9

Source: CNA (1999a).

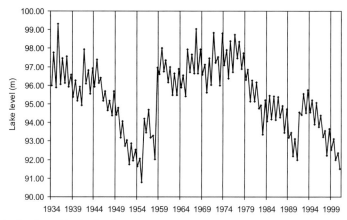

Fig. 8.2.   Lake Chapala water levels and basin developments.

	1930s	1940s	1950s	1960s	1970s	1980s	1990s
Rainfall (mm)[a]	670	648	685	757	740	668	720
Population (millions)[b]	2.5	3.0	3.6	4.5	5.9	8.7	11.0
Storage capacity dams (MCM)[c]	747	1,628	1,817	3,269	3,840	4,499	4,499
Irrigation (ha)[d]	n.a.	175,843	250,500	408,746	681,668	657,734	689,743
Lake inflow from Lerma (MCM)[e]	2,864	1,652	1,692	1,773	1,931	590	n.a.
Lake extractions (MCM)[e]	2,638	1,049	674	1,350	1,817	308	293

Sources for lake levels: de P. Sandoval (1994) and CNA (1991, 1992, 1993a, 1994, 1995, 1996, 1997, 1998, 1999b, 2000b).
[a]Decade average for de P. Sandoval (1994) for 1934–1949 and CNA/MW (1999) for 1950–1999.
[b]Population data for the end of the decade. Source: de P. Sandoval (1994) for extimates for 1939, 1949, 1959, 1969 and 1979, and CNA/MW (1999) for actual figures for 1989 and 1999.
[c]Constructed storage capacity at end of decade. Source: de P. Sandoval (1994) and CNA (1999e).
[d]Average actual irrigated area over the decade. Source: CNA (1999e).
[e]Decade average, excluding evaporation for de P. Sandoval (1994) and CNA (1991, 1992, 1993a, 1994, 1995, 1996, 1997, 1998, 1999b).

average to 90.8 m in 1954,[3] due to a prolonged drought combined with significant abstractions from the lake for hydroelectricity generation. During this period, around 200,000 ha were irrigated in the basin, mainly with surface water, and the constructed storage capacity in the basin was 1628 MCM. This period was the first time the basin headed towards closure as far as surface water is concerned. However, thanks to good rains towards the end of the 1950s, the lake recuperated, and levels fluctuated between 95.5 and 98.5 m from 1960 to 1979.

In 1979, a second period of decline set in, leading to basin closure in the mid-1980s. By this time, constructed storage capacity in the basin had increased to 4499 MCM and the average irrigated area had grown to around 670,000 ha, with a significant increase in groundwater irrigation. Although abstractions from the lake for hydropower generation had ceased, the combination of these factors resulted in declines in the lake level, from around 95 m at the start of 1980 to 92 m in 1990. After a modest recuperation in the early 1990s, lake levels declined again after 1994, and by February 2001 had fallen to 91.5 m, the lowest level measured since 1954, due to continued over-exploitation of surface and

---

[3]   Lake elevations are measured relative to a locally defined benchmark where 100 m is set as the high shoreline (de Anda *et al.*, 1998).

groundwater. It is unlikely that the lake will recover without exceptional runoff, as might be generated by a major hurricane. While the river basin was closing in water quantity terms, water quality also deteriorated severely, with increased effluent discharges and hardly any treatment of urban and industrial wastewater before 1989. Currently, the Lerma River and most of its tributaries are classified as contaminated. Two of its tributaries, the Turbio River and Guanajuato River, are classified as highly contaminated (CNA, 1999a).

8.2.1 Irrigated agriculture

The main water user in the basin is irrigation, depleting 59% of total available surface water and 79% of renewable groundwater (Table 8.1). Nine large-scale canal irrigation systems (termed irrigation districts in Mexico) command around 285,000 ha and some 16,000 farmer-managed and private irrigation systems (termed *unidades de riego* in Mexico) cover 510,000 ha. Twenty-seven reservoirs with a storage capacity of 250 MCM provide 235,000 ha in the irrigation districts with surface water, whereas around 1500 smaller reservoirs serve 180,000 ha in the *unidades*. An estimated 26,000 deep tubewells provide around 380,000 ha in the basin with groundwater, of which 47,000 ha are located in irrigation districts (CNA, 1993b; CNA/MW, 1999). In the irrigation districts, there are an estimated 88,000 water users (70,000 *ejidatarios* and 18,000 *pequeños propietarios*[4]) compared to 100,000 water users in the *unidades* (84,000 *ejidatarios* and 16,000 *pequeños propietarios*) (CNA/MW, 1999). Detailed data on cropping patterns and productivity for the whole basin are not available, although studies on selected irrigation systems are available (e.g. Kloezen and Garcés-Restrepo, 1998;

CNA/MW, 1999; Flores-López and Scott, 2000; Silva-Ochao, 2000). These studies show that the main crops irrigated in the basin are lucerne, wheat, sorghum and maize, whereas vegetables and fruits are increasing in importance.

In the early 1990s, the Mexican government transferred the government-managed irrigation districts to water users associations (WUAs) (cf. Rap *et al.*, 2004). Transfer was part of a major reform of the agricultural sector initiated at the highest level of government. Reform was driven by market-oriented economic and political imperatives and resulted in the following.

- Removal/reduction of direct and indirect subsidies to agricultural production;
- Privatization/elimination of public sector input supply and crop marketing bodies;
- Removal/reduction of tariffs and barriers to agricultural trade;
- Reform of the Mexican Constitution to permit the sale and renting out of *ejido* land.

The intent of these reforms was to stimulate economic growth through private investment in agriculture. The main objective of the Mexican irrigation management transfer (IMT) programme was to reduce public expenditure on irrigation by creating financially self-sufficient WUAs that would shoulder the full operation and maintenance (O&M) costs of the irrigation systems (Gorriz *et al.*, 1995; Johnson, 1997). Farmers were initially resistant to these changes, but have now generally come to accept them.

In the Lerma–Chapala basin, nine irrigation districts were transferred to farmer management in the early 1990s. WUAs now manage secondary canal units varying in size from 1500 to 30,000 ha. The WUAs were formed as legally recognized non-profit associations to which CNA granted 30-year concessions for the use of water and the

---

[4] *Ejidatarios* are members of *ejidos*, land reform communities created after the Mexican Revolution of 1910. Land holdings per *ejidatario* are typically less than 5 ha. *Pequeños propietarios* are private farmers with a limit on land ownership of 100 ha; however, holdings may be managed in much larger blocks, with nominal ownership in the hands of family members, friends and others.

irrigation infrastructure. In all the districts, CNA continues to manage the dams and headworks, and in most districts, it manages the main canals and delivers water in bulk to the WUAs as well. However, in the Alto Río Lerma irrigation district, a federation of WUAs has been formed to manage the main system (Kloezen, 2000).

Research carried out by Kloezen *et al.* (1997) shows that the WUAs in the Alto Río Lerma irrigation district have been effective in improving the provision of services and recovering costs from users.[5] More recent work in the district raises questions about the WUAs' long-term sustainability (Monsalvo, 1999; Kloezen, 2000). It is important to note that IMT was externally imposed and occurred at a time when the Lerma–Chapala basin was already closed. This has placed a double strain on the newly created WUAs. On the one hand, they have had to learn how to manage sizeable secondary canal units, at which they have been relatively successful, and how to interact with the CNA. On the other hand, they need to organize themselves by establishing relations with WUAs in other irrigation districts to ensure their voice is heard in river basin management discussions.

The management structures in the *unidades* are much more diverse, and may consist of informal WUAs, government recognized WUAs, water judges, pump groups or commercial management (Silva-Ochoa, 2000). As state intervention in the *unidades* has been piecemeal in comparison to the districts, and has usually only consisted of assistance in construction and the concessioning of water rights, government control over water use in the *unidades* is much weaker than in the irrigation districts. As a corollary, the representation of *unidades* in basin-wide decision-making forums is weak.

8.2.2 Urban water supply

Domestic water supply in the basin mainly depends on groundwater (95%), with total consumptive use at 791 MCM. In addition, water is transferred out of the basin to provide Guadalajara (237 MCM surface water) and Mexico City (323 MCM groundwater) with urban water. The population in the basin has increased significantly, doubling from 2.1 million inhabitants in 1930 to 4.5 million in 1970 and then more than doubling again to 11 million in 2000 (CNA/MW, 1999). The population of the basin will again double in the next 30 years if the population growth rate of 2.16% remains the same (CNA/MW, 1999). Besides a fivefold increase in population in the past 70 years, the basin's population has become strongly urbanized. Population in the seven largest cities in the basin increased from 12.7% to 40.9% of the basin's total population between 1930 and 2000, while the rural population dropped from above 75% to less than 25% during the same period (CNA/MW, 1999). Population growth has led to increasing pressures on the basin's water resources. Scott *et al.* (2001) project that urban water demand in the medium term will increase by some 4.1% per year.

Starting in 1983, domestic water supply, wastewater collection, and, more recently, the operational costs of wastewater treatment, were decentralized to municipalities. The creation of water utilities has been promoted to separate these activities from other municipal responsibilities. However, according to CNA (1999d, p. 8) 'most of the water utilities have a poor performance and need to be greatly improved to achieve technical and economical sufficiency.'

---

[5]   Based on an extensive survey of farmers' perceptions on changes in irrigation management after transfer, Kloezen *et al.* (1997) report that 36% of the farmers perceived that water adequacy at field level had improved with IMT, 26% perceived no change and 23% reported it had become worse. According to 64% of the *ejidatarios* and 47% of the private farmers the condition of the irrigation network had improved, while 54% of the *ejidatarios* and 38% of the private farmers stated the drainage network had improved. Cost recovery went up from 50% to more than 100%.

8.2.3 Industry

Although industry only uses a small amount of water (278 MCM or 3% of annual renewable water in the basin), it generates 35% of Mexico's industrial GNP and pays around US$42 million in water taxes[6] to the federal government (CNA/MW, 1999). The 6400 registered industrial firms in the basin are a major source of water pollution (figures are not available), although officially they must have a permit from CNA indicating effluent standards to discharge wastewater. Fees for discharging wastewater above effluent standards were established in 1991 and updated annually. According to Tortajada (1999), most polluters do not pay these fees, while industries and cities that have submitted wastewater treatment proposals to CNA are exempted from payment.

8.3 The Hydro/Institutional Landscape in the Lerma–Chapala Basin

A watershed year for water management in Mexico was 1989. Whereas the previous 100 years were characterized by increasing federal control over water, since 1989 decentralization has been the norm. Currently states, municipalities and water users have a larger say in water management decision making. The current distribution of competencies in water management in the Lerma–Chapala basin is set out below.

8.3.1 Water rights

In Mexico, surface water is defined in the Constitution as national property placed in the trust of the federal government. As the trustee of the nation's water, the federal government has the right to concession surface water-use rights to users for periods ranging from 5 to 50 years (Kloezen, 1998). In irrigation districts and *unidades*, concessions are granted to WUAs and not individual water users. The concession titles set out the volumes of water concession holders are entitled to, although CNA may adjust the quantity each receives annually to reflect water availability, with priority being given to domestic water users (CNA, 1999b). Thus for allocating surface water, Mexico follows the proportional appropriation doctrine and, in theory, all concession holders share proportionally in any shortages or surpluses of water.[7] Once issued, water concessions need to be recorded in the Public Register of Water Rights maintained by CNA. After registration, the concessions become fully tradeable within river basins, although the CNA needs to be notified of trades and needs to approve them (Kloezen, 1998).

The situation surrounding groundwater is more complex, as the Constitution does not define it as national property but rather states that overlying landowners may bring groundwater to the surface as long as this does not affect other users. In 1946, the Constitution was amended to enable the federal government to intervene in aquifers in overdraft by issuing pump permits or declaring that new pumps may not be installed. Based on a ruling of the Supreme Court in 1983, groundwater is now considered national property, although this is not reflected in the Constitution or the 1992 water law (Palacios and Martínez, 1999). Groundwater concessions in Mexico are granted on a volumetric basis, with a maximum extraction or pumping rate specified.

---

[6]   Under the Federal Law on Excise Taxes, updated annually, water users have to pay a tax for the benefits derived from using water, because it is national property. For agriculture a zero rate was established; farmers only have to pay irrigation service fees to WUAs (Tortajada, 1999).

[7]   This contrasts with the prior appropriation system, where first rights have seniority.

8.3.2 Water management organizations and stakeholders

Numerous stakeholders are involved in water management in the Lerma–Chapala basin. The government agency responsible for water management in the basin is the CNA, a semi-autonomous federal agency falling under the Ministry of Environment. Created in January 1989, the CNA is charged with defining water policy, granting water concessions and wastewater discharge permits, establishing norms for water use and water quality and integrating regional and national water management plans. The aim of unifying all government responsibilities related to water in the CNA was to create the necessary conditions for moving towards sustainable water management (CNA, 1999d). To complement this move, a modern and comprehensive water law was promulgated in 1992 (CNA, 1999c). This law defines an integral approach for managing surface and groundwater in the context of river basins, which it considers the ideal unit for water management. It also promotes decentralization, stakeholder participation, better control over water withdrawals and wastewater discharges, and full-cost pricing.

To discharge its mandate, CNA consists of four levels: federal headquarters, regional offices, state offices and irrigation district offices. Specific responsibilities of the federal headquarters include the following:

- Manage the nation's water and act as its custodian;
- Formulate and update the National Water Plan and ensure its execution;
- Maintain the Public Register of Water Rights;
- Monitor water taxes payment and send reports to fiscal authorities;
- Facilitate the resolution of, or arbitrate in, water conflicts;
- Promote conservation and efficient use of water;
- Support the development of urban and rural domestic water supply and drainage and sanitation networks, including the treatment and reuse of wastewater;

- Support the development and management of irrigation and drainage systems and storm and flood protection works.

To facilitate river basin management and interaction with stakeholders, CNA has divided the country into 13 hydrologic regions based on river basin boundaries and established an office in each region. These regional offices have been delegated responsibilities from the national level and are relatively autonomous. Their main responsibilities include the following:

- Organize and manage CNA's actions at the regional level concerning the planning, execution and evaluation of the Regional Water Plan;
- Integrate and validate requests for water concessions (users), allocations (municipalities) and permits (groundwater and wastewater), issuing those that fall in its competency and forwarding to HQ those that do not;
- Supervise the Public Register for Water Rights offices at the state level, and consolidate and send to HQ all information necessary to keep the Register updated at the national level;
- Assist CNA state offices in their collection of water taxes from users;
- Integrate and update programmes for the operation, maintenance and rehabilitation of irrigation districts and water treatment plants;
- Supervise and assist in the operation of climate and river flow measuring networks;
- Enforce legal standards concerning water pollution and wastewater discharges.

Responsibilities for water management at the state level are more diffuse. CNA has established offices in Mexico's 31 states that function under the supervision of the regional offices. The role of state governments, as opposed to the federal government, in water management has been limited to regulating municipal water utilities and supporting utilities showing poor technical and economic performance. As part of the 'new federalism' policy during the Zedillo administration (1995–2000), the federal government promoted delegation of responsibilities and programmes to the states, but, notably, not financial resources. Although the federal government has encouraged the modification of state laws to promote the participation of state governments in water management through the creation of State Water Commissions, the response has been lukewarm. This is not the case in Guanajuato, where CEAG (*Comisión Estatal del Agua de Guanajuato*; Guanajuato State Water Commission) has taken on its new role with vigour (Guerrero-Reynoso, 2000). The relationship between the State Water Commissions and the CNA still needs to be defined.

At the irrigation district level, CNA and WUAs share responsibility for water management. According to the 1992 water law and the WUAs' concession titles, CNA remains the authority in the irrigation districts. The CNA Irrigation District Office retains the following responsibilities and powers:

- Notify the WUAs on the volume of surface water they have been allocated for the coming year;
- Operate and maintain the dams and headworks of the irrigation district and also the main system if a federation of WUAs has not been established;
- Approve irrigation service fee levels, determined by the WUAs according to the procedures outlined by the CNA in the concession title;
- Establish and periodically revise the instructions for the operation and maintenance of the secondary canal units managed by WUAs;
- Approve the WUAs annual maintenance plan and ensure that it is carried out satisfactorily;
- Participate in the General Assembly of the WUAs with the right to speak but not to vote;
- Cancel or refuse to renew concession titles if WUAs perform unsatisfactory.

Based on the concession granted to them by the CNA, WUAs legally assume the responsibility to operate, maintain and administer their secondary canal unit. Their specific responsibilities to the CNA are the following:

- Develop and enforce bylaws that detail procedures for water distribution, system maintenance and investment, cost recovery, and dealing with complaints and sanctions;
- Collect irrigation service fees that fully cover the O&M and administration costs of the WUA;
- Pay CNA a percentage of the revenues from fee collection for CNA services related to O&M of the dams, headworks and main canal system;
- Prepare annual operation and maintenance plans and budgets and send to CNA for approval.

At the river basin level, an important innovation has been the creation of river basin councils. The water law stipulates that stakeholder participation is mandatory in water management at the river basin level. To this end river basin councils, defined in the water law as coordinating and consensus-building bodies between the CNA, federal, state and municipal governments, and water user representatives (CNA, 1999c), have been established by CNA in 25 river basins (CNA, 2000a). The stated goal of the councils is to foster the integral management of water in their respective river basins through proposing and promoting programmes to improve water management, develop hydraulic infrastructure and preserve the basin's resources. Formally, the river basin councils have little decision-making power, as CNA remains responsible for water concessions, the collection of water taxes and water investment programmes. The role of the councils is to assist CNA in the execution of its vested powers and to ensure that CNA takes stakeholders' opinions into account (CNA, 2000a).

Mexico's first river basin council was established in the Lerma–Chapala basin in response to the dropping level of Lake Chapala in the 1980s and the severe contamination of the Lerma River (Mestre, 1997). In

April 1989, the federal government and the five state governments signed a coordination agreement to improve water management in the basin by: (i) allocating surface and groundwater fairly among users and regulating water use; (ii) improving water quality by treating effluents; (iii) increasing water-use efficiency; and (iv) conserving the river basin ecosystem and watersheds. Based on this agreement, a formal Consultative Council was formed in September 1989 to follow up on these objectives. Based on the 1992 water law the Consultative Council became the Lerma–Chapala River Basin Council on 28 January, 1993.

In the past 10 years, the River Basin Council has been in flux, and only in August 2000 was its structure formalized (CNA, 2000a). It now consists of a Governing Council, a Monitoring and Evaluation Group (MEG), a Basin Level User Assembly and Special Working Groups, while the CNA's regional office forms the Council's secretariat (Fig. 8.3). The Governing Council is chaired by CNA, while user representatives from six sectors (agriculture, fisheries, services, industry, livestock and urban) and the governors of the five states falling in the basin are its members, yielding a total of 12

members. Although agriculture uses 68% of the annual renewable water in the basin, it only has one vote out of 12 on the Council. The decision-making body of the River Basin Council is the MEG, which is a carbon copy of the Governing Council except that state governors send representatives in their stead, while CNA is represented by the head of its regional office. The MEG meets on a regular basis to prepare and convene Council meetings and more importantly to draft agreements to be signed at formal Council meetings.

The structure of the River Basin Council is complemented by a stepped form of user representation, consisting of water user committees for the six water use sectors represented on the Council. These committees can be formed at the regional, state or local level, building on already existing WUAs or other legally recognized water management groups where possible. The water user committees form the Basin Level User Assembly, which elects the six user representatives on the Council. In addition, forums at the sub-basin level, such as watershed commissions and aquifer management councils, form part of the structure of the River Basin Council (Fig. 8.3).

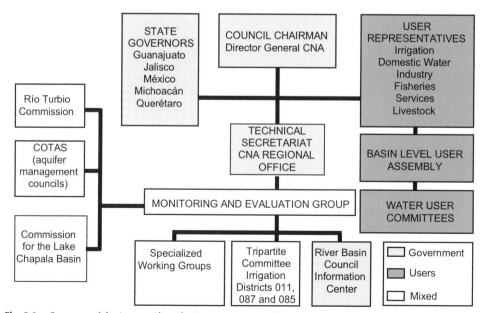

Fig. 8.3. Structure of the Lerma–Chapala River Basin Council.

8.4 Essential Functions for River Basin Management

As set out in Chapter 1, a set of essential functions has been posited to analyse the institutional arrangements for river basin management. A list of these functions as they apply in the Lerma–Chapala basin, crossed with the key actors identified in the previous section, is shown in Table 8.2. Cells are marked to indicate if an actor is active in a particular functional area, while action verbs such as Execute (E), Supervise (S), Advise (A), Authorize (Aut), Regulate (Reg) and Represent (Rep) are used to describe the nature of an actor's activities or responsibilities. The table, together with the description of the actors, gives an indication of which essential functions are accounted for and who is involved in their execution. In itself, this is useful to identify whether certain essential functions are not being executed. A number of interesting points emerge from Table 8.2:

- Most of the essential functions relating to surface water are covered.
- The withdrawal and distribution of groundwater is weakly regulated, and the construction of new facilities continues although this has been prohibited.
- Monitoring and ensuring the quality of primary water sources are relatively weak functions. Stopping groundwater pollution is a function that is not covered at all.
- The monitoring of wastewater quality and enforcing wastewater discharge norms are also functions that receive little attention, while the allocation and distribution of wastewater in the basin is not regulated at all and left to actors on the ground.
- Industries and non-governmental organizations (NGOs) are hardly involved in water management in the basin.

Another, perhaps more interesting, question is how efficiently the functions are executed and if they are effective. To evaluate how well the functions are being performed, an evaluation matrix was developed and applied to the Lerma–Chapala basin. This matrix consisted of eight variables for each essential function, four related to effectiveness and four related to efficiency. For each variable, three questions were formulated that could be answered on a scale of 1 to 5, giving a maximum score of 60 points for effectiveness and 60 points for efficiency per function. The scores per function can be set out against each other, with effectiveness on the x-axis and efficiency on the y-axis, yielding a graphical representation of how well the functions are executed (Fig. 8.4). The scores presented in Fig. 8.4 are the averages of the responses given by 15 key water management stakeholders in the basin. Although a drawback of the evaluation matrix is that it did not distinguish between surface and groundwater or between primary and derivative water sources, it is presented here as it reflects overall impressions of key water management actors in the Lerma–Chapala basin regarding the execution of the essential functions in their basin. An interesting point that emerges from Fig. 8.4 is that none of the essential functions are executed efficiently according to key water managers in the basin. What also stands out is that, apart from protecting the environment, the effectiveness of the execution of the essential tasks is moderate. The next section presents more detail on the overarching issues facing the basin.

8.5 Overarching Issues

Through steady changes in the institutional arrangements in the Lerma–Chapala basin in recent years, progress has been made towards improved water management. This progress is significant, in light of the complicated transition from highly centralized water management to one in which states, municipalities and water users have a larger say. None the less, from a water perspective the Lerma–Chapala basin is still in crisis. The efforts of the Council over the past 10 years need to be redoubled to tackle

Table 8.2. Key actors and essential basin management functions in the Lerma–Chapala basin.

Key actors	Surface water									Groundwater							Derivative water					
	Plan (basin-level)	Allocate water	Distribute water	Constuct facilities	Maintain facilities	Monitor quality	Ensure quality	Protect against flooding	Protect ecology	Plan (basin-level)	Allocate water	Withdraw/distribute water	Construct facilities	Maintain facilities	Monitor quality	Ensure quality	Plan (basin-level)	Allocate/distribute	Construct facilities	Operate/maintain facilities	Monitor quality	Enforce quality
Ministry of Environment	Reg	Aut		Reg		Reg	Reg		Reg		Reg										Reg	Reg
CNA National Headquarters	Aut	Aut		Aut	S	S	S	S	S	S	Aut		Aut				S		Aut		S	S
CNA Regional Office	E	E/S	S	S	S	S/E	S/E	S	E	S	S		S		S		S		S/A	S	S/E	E
River Basin Council	Rep	Rep		Rep							Rep						Rep		Rep	A		
CNA State Office	A	E	S	E	S	E	E	E	A	A	E	S	S		E				E	A	E	E
State Water Commissions	E/A	A	A	E/A	S/A			S/A	A	A	A		E	A	E		A		S/A	A	E	
CNA Irrigation District Office			E	E	Aut																	
WUAs Irrigation Districts	Rep		E	Rep	E							E						Aut				
WUAs Irrigation Units			E		E							E										
Aquifer Management Councils (COTAS)	Rep								A	A	Rep		A		E							
Municipal Water Supply Utilities			E	E	E			E				E	E	E	E			Aut	E	E	E	
Industries				E								E	E	E					E	E		
NGOs	A								A													
Irrigators			E									E	E	E				E				

E, Execute; S, Supervise; A, Advise; Aut, Authorize; Reg, Regulate; Rep, Represent; CNA, Comisión Nacional del Agua; WUA, water users' association; NGO, non-governmental organization.

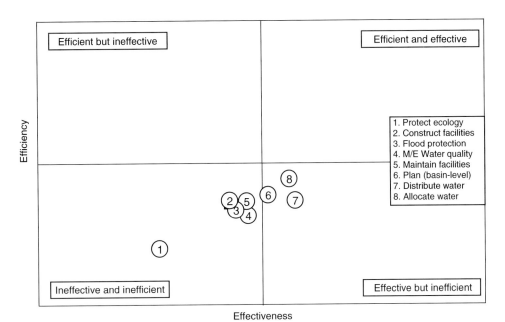

Fig. 8.4. Effectiveness and efficiency of essential functions in the Lerma–Chapala basin.

the significant challenges lying ahead of it. Three challenges stand out, namely the allocation of surface water and groundwater, and the management of derivative water.

8.5.1 Surface water allocation

To allocate surface water in the basin, the governors of the five states in the basin and the federal government signed a treaty in August 1991 (CCCLC, 1991). An important objective of the treaty is to maintain adequate water levels in Lake Chapala and to ensure Guadalajara's domestic water supply. To preserve Lake Chapala, the treaty sets out three allocation policies, namely *critical*, *average* and *abundant*, based on whether the volume of water in the lake is less than 3300 MCM, between 3300 and 6000 MCM, and more than 6000 MCM, respectively. Each year the Council verifies the volume stored in Lake Chapala to determine the allocation policy to be followed for the next year. For each allocation policy, formulas have been drawn up to calculate

water allocations to the irrigation systems in the basin, based on the surface runoff generated in each of the five states in the previous year. Table 8.3 indicates how this works for the Alto Río Lerma irrigation district. Based on extensive modelling of these formulas, it was concluded that the resulting water allocation would not impinge on the 1440 MCM needed by Lake Chapala for evaporation. Thus, as shown in Table 8.3, if the surface runoff generated is below a certain threshold, a fixed volume is deducted from the irrigation district's allocation so that this can be passed on to the lake.

Since 1991, the MEG of the Council has met each year and has applied the water allocation rules set out in the treaty. Figure 8.5 sets out the volumes of water allocated and used from 1992 to 2000, as well as the volume of water stored in Lake Chapala. This shows that the 1991 treaty has been enforced, as actual use has never been higher than the allocated values. A caveat here is that only the extractions by irrigation districts are accurately measured, thus actual withdrawals may have been higher as the amount of water going to the *unidades de riego* is unknown.

Table 8.3.   Water allocation principles for the Alto Río Lerma irrigation district.

Lake Chapala volume	Surface runoff generated (SRG) in the State of Guanajuato (MCM)	Volume allocated (VA) to irrigation district (MCM)
Critical	if SRG between 280 and 1260	then VA = 94.2% of SRG–262.8
	if SRG > 1260	then VA = 924
Average	if SRG between 144 and 1125	then VA = 94.2% of SRG–135.6
	if SRG between 1125 and 1400	then VA = 924
	if SRG > 1400	then VA = 955
Abundant	if SRG between 19 and 1000	then VA = 94.2% of SRG–17.9
	if SRG between 1000 and 1200	then VA = 924
	if SRG > 1200	then VA = 955

Source: CCCLC (1991).

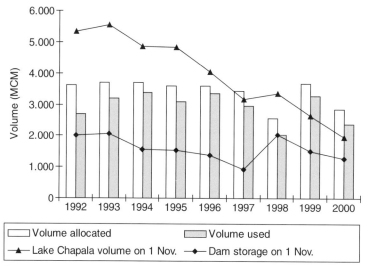

Fig. 8.5.   Surface water allocated and used in the Lerma–Chapala basin. Sources: CNA (1991, 1992, 1993a, 1994, 1995, 1996, 1997, 1998, 1999b, 2000b).

Despite the apparent compliance with the surface water treaty, Lake Chapala's volume has halved in the past 8 years. This is so, in part, because the treaty takes the surface runoff generated in the previous year to determine water allocations. In 1997 rainfall was only 645 mm (against the average annual value of 712 mm) and dam storage (used here as a proxy of surface runoff) was consequently low. Combined with a lake volume below 3300 MCM, the critical allocation policy was followed for 1998, leading to the lowest allocations since the treaty was signed. However, rainfall in 1998 was exceptionally good, at 810 mm some 100 mm above average, leading to a recuperation of the volume of water stored behind dams and a slight increase in the volume of Lake Chapala to 3361 MCM. As a result, the average allocation policy was followed for 1999 and 3664 MCM were allocated to water users, the highest level since the signing of the treaty. Unfortunately, rainfall in 1999 was a historic low of 494 mm. These two factors resulted in Lake Chapala dropping to its lowest level since the signing of the treaty and point to inadequate provisions in the treaty for inter-annual planning of water availability and dealing with contingencies.

The members of the Council have recognized the shortcomings of the surface water treaty and in 1999 decided to revise the treaty, as it was clear that it was not rescuing Lake Chapala. In 1999 and 2000, detailed studies were carried out with hydrological data from 1945 to 1997 (an improvement over the 1950–1979 data used for the previous treaty) to develop a new model for calculating surface runoff (CNA, 1999e). The Council signed the amendment of the 1991 surface treaty on 24 August, 2000 (Consejo de Cuenca Lerma–Chapala, 2000). However, various states in the basin feel that they did not have sufficient input in the design of the surface runoff model and that CNA imposed the treaty on them. In addition, consultation with water users concerning the new treaty has been minimal, although user representatives on the Council voted in favour. Although the signing of the new treaty shows the adaptability of the Council and the commitment of its members to construe a water allocation policy that meets urban and agricultural needs while safeguarding the environment, the process by which it was arrived at is contested.

An issue that the Council has not yet started to consider is how to compensate farmers for water transferred out of agriculture for urban and environmental demands. Scott *et al.* (2001) calculate that the benefits forgone for farmers in the Alto Río Lerma irrigation district as a result of a reduced water allocation to the district for 1999/2000 amounted to US$14 million. Although sufficient water was stored in the district's reservoirs to cover its full allocation (955 MCM) the district was allocated only 648.2 MCM under the treaty, due to the critically low volume of water in Lake Chapala and the minimal surface runoff generated in Guanajuato in 1999. To shore up water levels in Lake Chapala, the CNA released 240 MCM from the district's main reservoir into the Lerma River in October 1999, thereby transferring surface water from the agricultural sector to the urban and environmental sector.

The reduced allocation to the Alto Río Lerma irrigation district in 1999 resulted in some 20,000 ha out of a total of 77,000 ha not being irrigated with surface water in the winter season. Four of the 11 WUAs in the district went without irrigation altogether, whereas in the other WUAs, irrigation was restricted to a maximum of 3 ha per landowner. For better off farmers who could switch to groundwater, this was not too problematic, but for many poorer farmers who mainly rely on surface water, the results were disastrous. In addition, many poor farmers who traditionally pumped return flows from the Lerma River were hard hit as the use of this precarious source of water was prohibited and enforced through army patrols along the river. The surface water allocations for all the irrigation districts in the basin for the 2000/01 winter season were so low that the WUAs decided to let 200,000 ha (of a total of 235,000 ha in the irrigation districts normally irrigated with surface water) lie idle. Once again, poor farmers were hardest hit. On the other hand, Lake Chapala has dropped to its lowest levels in 50 years, and environmental NGOs and the Jalisco state government are demanding the transfer of water stored behind the dams of the irrigation districts to the lake.

The lack of water in the past 2 years has galvanized WUA leaders to take action. In May 2000, the presidents of WUAs located in Jalisco, Guanajuato and Michoacán met each other for the first time to discuss ways to strengthen their representation in the River Basin Council. Until then, WUAs of a particular irrigation district had dealt only with the CNA, and there were no horizontal linkages between WUAs from different districts. In 2001, WUAs from irrigation districts in Querétaro and Mexico joined the discussions, and the combined WUAs established a new working group in the River Basin Council focusing on agriculture. They have vowed that not a single drop of water will be passed to Lake Chapala and have threatened civil disobedience if CNA transfers water as in October 1999. It is clear that a serious conflict is brewing in the basin surrounding surface water and that the Council needs to act fast to come up with a workable solution. What is more worrisome is that the lack of surface water has led many farmers to increase their use of groundwater, while aquifers are already in perilous decline.

8.5.2 Groundwater management

The serious overdraft of the basin's aquifers is arguably a more pressing issue than surface water allocation. Although the Council signed a coordination agreement to regulate groundwater extraction in the basin in 1993, progress on the ground has been slow (CCCLC, 1993). A key problem is that the Council, through the CNA, does not physically control the water extraction infrastructure (the wells), as it does in the case of surface water (the dams). Although the Constitution mandates the federal government to intervene in aquifers in overdraft by placing them under *veda*, prohibiting sinking of new wells without permission from the federal government, the experience with *vedas* has been disappointing (Arreguín, 1998). The reality of groundwater extraction in Guanajuato clearly shows how groundwater regulation by the federal government has run aground. According to Vázquez (1999) 10 *vedas* were issued in Guanajuato between 1948 and 1964 prohibiting the drilling of new wells in large parts of the state, and in 1983, the remainder of the state was placed under *veda*. Notwithstanding these legal restrictions, the number of wells in Guanajuato increased from approximately 2000 in 1958 to 16,500 in 1997 (Guerrero-Reynoso, 1998).

Based on the recognition that *vedas* have not worked and to counter the continued depletion of groundwater in the basin, CNA started promoting the formation of Comités Técnicos de Aguas Subterráneas (COTAS; aquifer management councils) in selected aquifers in 1995, as an outgrowth of the 1993 agreement. Through the establishment of COTAS, which fall under the River Basin Council, the CNA is seeking to stimulate the organized interaction of aquifer users with the aim to establish mutual agreements for reversing groundwater depletion, in keeping with Article 76 of the water law regulations (CNA, 1999c). Based on recent developments in the State of Guanajuato, where CEAG enthusiastically promoted the creation of COTAS (Wester *et al.*, 1999; Guerrero-Reynoso, 2000), the structure of the COTAS has been defined at the national level in the rules and regulations for river basin councils (CNA, 2000a). In these rules, the COTAS are defined as fully fledged user organizations, whose membership consists of all the water users of an aquifer. They are to serve as mechanisms for reaching agreement on aquifer management taking into consideration the needs of the various sectors using groundwater (CNA, 2000a).

To date, 14 COTAS have been formed in the basin. However, none of them has yet started to devise ways to reduce groundwater extraction. Considering that some 380,000 ha in the basin are irrigated with groundwater and that industrial and domestic uses depend almost entirely on groundwater, it is fair to say that groundwater is the strategic resource in the basin, particularly from a water productivity perspective. The long-term consequences of its continued depletion easily overshadow those of Lake Chapala drying up. Although the COTAS are a timely institutional response to the pressing need for innovative approaches to managing aquifers in the basin, it remains to be seen if they will succeed in reducing aquifer over-exploitation. Current discussions in the COTAS focus on installing sprinkle and drip irrigation systems to save groundwater, but the tough issue of how to reach agreement on an across the board reduction in pumping has not yet been broached. In addition, new pumps continue to be installed and regularized through extra-legal means. The reluctance of the government to impose strict pumping limits and the continued race to the pumphouse by farmers bode ill for the COTAS.

8.5.3 Management of derivative water

The effects of the new institutional arrangements in the basin on surface and groundwater management were briefly described above. The picture that emerges is that, although some progress has been made, the basin's primary water sources are still seriously overdrawn, to the particular detriment of poorer farmers. For them, derivative water is becoming a critical, and frequently the only, available source of water (cf. Buechler and Scott, 2000). In 1993, the municipal wastewater flow generated in the basin was estimated at 12,700 l/s, and by 1997, this had risen to 17,000 l/s, which is equal to around 536 MCM per year (Mestre, 1997). The return flows from agriculture are not measured, but the fact that hardly any water flows into Lake Chapala suggests that, whatever the magnitude of the return flows, they are used upstream. The extent of wastewater irrigation in the basin has not been accurately assessed, but estimates range from 20,000 to 40,000 ha (CNA/MW, 1999). The use of return flows and wastewater is currently a free-for-all and is not formally regulated by the institutional arrangements for water management in the basin. However, as Buechler and Scott (2000) outline, farmers at times need to obtain informal permission from municipalities or WUAs to use wastewater. These claims are being threatened by the construction of wastewater treatment plants and the *de facto* reallocation of treated water to other uses.

In 1989, the River Basin Council launched an investment programme to clean the Lerma River through the planned construction of 48 treatment plants with a capacity of 3700 l/s. Priority was given to cities with more than 10,000 inhabitants that discharged directly into the Lerma River or Lake Chapala. This was essential, as up to 1989 hardly any of the municipal and industrial wastewater in the basin was treated. As a result, large stretches of the Lerma River were heavily contaminated, as were two of its tributaries, the Turbio and Querétaro Rivers, while Lake Chapala was classified as contaminated. On 28 January, 1993, the river basin council agreed on a second water treatment programme, entailing the construction of 52 new plants and the enlargement of five existing ones, with a combined treatment capacity of 10,835 l/s (Mestre, 1997). It was foreseen that the two investment programmes would create a total treatment capacity of 14,561 l/s, representing 85% of municipal wastewater generated in the basin. At the end of 1999, 48 plants had been constructed

with a treatment capacity of 6037 l/s, while six additional large plants are still under construction, with a capacity of 3513 l/s (CNA, 1999a). Of the installed capacity of 6037 l/s, only 3155 l/s provides full treatment. Although the River Basin Council has not succeeded in attaining the ambitious goals it set for itself, the installed capacity of 6037 l/s, up from 0 l/s in 1989, is an achievement in itself. The Council readily admits that more needs to be done, although the large investments required and the lack of financial resources at the basin level makes this difficult (CNA, 1999a).

Scott *et al.* (2000) place the value of wastewater irrigation for 140 ha of irrigated land that depend on the City of Guanajuato for their water at some US$252,000 per year. As part of the Council's water treatment programme, Guanajuato City is in the process of contracting for an activated sludge wastewater treatment plant and plans to sell the treated water to commercial interests. As a result, farmers will lose their access to wastewater as well as the nutrient benefits of that water. This raises the question how a *de facto* right to derivative water should be addressed when water treatment redirects wastewater outflows. Although the CNA, the River Basin Council and the State Water Commissions are actively pursuing the construction of wastewater treatment plants, the allocation and distribution of derivative water has received no attention. In closing river basins, where primary water sources have already been captured by the better off and renegotiating water rights is extremely complicated, derivative water rapidly becomes the poor farmer's last resort and should be recognized as such.

8.6 Conclusions

This chapter has explored the dynamic interplay between trends in water use and institutional changes for water management in the Lerma–Chapala basin. Drought and water shortages between 1945 and 1954 resulted in a doubling of the reservoir capacity within the basin. Pressures on available primary water, both surface and ground, continued to increase with the irrigated area increasing almost fourfold up to 1989 and the population increasing almost threefold. After primary water resources had become over-committed, institutional reforms were introduced from 1989 onwards devolving responsibility for water management in irrigation systems to water users and initiating participatory water management bodies at the basin level for high-level decision making on water allocation.

A central component in this reform has been the 1992 water law, though of equal importance have been the institutional capability to put the law into practice and institutions' ability to adapt to a dynamic situation with additional measures for controlling and managing available surface and groundwater resources. The assessment of the institutional arrangements for water management in the Lerma–Chapala basin brings out clearly the need for coordinating mechanisms at the basin level in river basins facing closure. Through the Lerma–Chapala River Basin Council, progress has been made towards improved water management in the basin. This progress is encouraging, in light of the complicated transition from a highly centralized management of water resources to one in which states, municipalities and water users have a larger say. Though the institutional measures have had a significant impact in restructuring water management, the basin is still in crisis, with the level of Lake Chapala in perilous decline and aquifers running dry. The efforts of the River Basin Council over the past ten years will need to be redoubled to tackle the significant challenges lying ahead of it.

Although surface water allocation mechanisms are working, and the revision of the 1991 treaty may lead to increased inflows to Lake Chapala, compensating farmers for water transferred out of agriculture needs to be considered. In closed basins, inter-sectoral transfers are inevitable and it will invariably be the irrigation sector that will need to cede water. A key institutional challenge in closed river basins is dealing with these transfers in a just and equitable manner, such that the

sustainability of locally managed irrigation is ensured. The Lerma–Chapala River Basin Council could be a forum for drawing up and enforcing compensation mechanisms for surface water transferred out of agriculture, but to date it has been reluctant to consider this issue due to the costs involved. To be credible and effective partners in basin management, WUAs and farmers need to make concerted efforts to reduce water use, for which there is real scope, and to join forces to argue their case more strongly in the River Basin Council.

A much more serious challenge that the Council and other water management stakeholders in the basin need to deal with urgently is the overdraft of the basin's aquifers. The aquifer management councils are a step in the right direction, but their role in groundwater management should go beyond mere consultation. Bundling extraction rights in an aquifer and concessioning this to a COTAS is feasible under the Mexican water law and should be seriously considered. Placing aquifer management in the hands of the aquifer users, under the supervision of the River Basin Council, State Water Commissions and the CNA, shows more promise of reducing extractions than the current system of *vedas* and federal regulation.

With basin closure, poorer farmers in particular are losing, or have lost, their access to primary water due to reductions in surface irrigation and increased costs for groundwater irrigation. Hence derivative water, both wastewater and agricultural drainage effluent, is increasingly becoming a critical resource for poor farmers. However, its allocation in the basin is currently a free-for-all. Although progress has been made in constructing wastewater treatment plants, the management of derivative water has received little attention. More generally, meeting the water needs of poor people, and including poor women and men at all levels of water management decision making, is not a priority of the Council, nor is it a strong feature of the larger set of institutional arrangements for water management in Mexico. The Council needs to seriously consider how to safeguard and improve the access of the poor to water, and how to combat the current *de facto* concentration of water rights in the hands of the few. Overall, a coherent approach for tackling the continued pressures on the basin's primary water sources, the significant re-use of wastewater (derivative water) and the deterioration of water quality needs to be developed.

References

Arreguín, J. (1998) *Aportes a la historia de la geohidrología en México, 1890–1995.* CIESAS-Asociación Geohidrológica Mexicana, Mexico City.

Buechler, S.J. and Scott, C.A. (2000) 'Para nosotros, esta agua es vida.' El riego en condiciones adversas: Los usuarios de aguas residuales en Irapuato, México. In: Scott, C.A., Wester, P. and Marañón-Pimentel, B. (eds) *Asignación, productividad y manejo de recursos hídricos en cuencas.* International Water Management Institute (IWMI), Serie Latinoamericana No. 20. IWMI, Mexico City.

CNA (Comisión Nacional del Agua) (1991, 1992, 1993a, 1994, 1995, 1996, 1997, 1998, 1999b, 2000b) Volumenes maximos de extracción de agua superficial para los sistemas de usuarios de la cuenca Lerma-Chapala. *Boletin No. 1–10.* CNA, Mexico City.

CNA (1993b) *Plan maestro de la cuenca Lerma-Chapala. Documento de Referencia.* CNA, Mexico City.

CNA (1999a) *El consejo de cuenca Lerma–Chapala 1989–1999. 10 años de trabajo en favor de la gestión integral y manejo sustentable del agua y de los recursos naturales de la cuenca.* CNA, Guadalajara.

CNA (1999c) *Ley de aguas nacionales y su reglamento.* CNA, Mexico City.

CNA (1999d) *Regional Vision North America (Mexico).* CNA, Mexico City.

CNA (1999e) *Estudio de disponibilidad y balance hidráulico de aguas superficiales de la cuenca Lerma-Chapala y cuencas cerradas de Pátzcuaro.* CNA, Guadalajara.

CNA (2000a) *Reglas de organización y funcionamiento de los consejos de cuenca.* CNA, Mexico City.

CNA/MW (1999) *Proyecto lineamientos estratégicos para el desarrollo hidráulico de la Región Lerma–Santiago–Pacifico. Diagnostico regional.* CNA/Montgomery Watson, Guadalajara.

CCCLC (Consejo Consultivo de Evaluación y Seguimiento del Programa de Ordenación y Saneamiento de la Cuenca Lerma-Chapala). (1991) Acuerdo de coordinación de aguas superficiales. *Colección Lerma-Chapala*, vol. 1, no. 5. CNA, Queretaro.

CCCLC (1993) *Propuesta de acciones para reglamentar la distribución, uso y aprovechamiento de las aguas del subsuelo en los acuiferos que se ubican en la cuenca Lerma-Chapala.* CNA, Queretaro.

Consejo de Cuenca Lerma-Chapala (2000) *Actualización de las bases y procedimientos para el cálculo de disponibilidad y distribución de las aguas superficiales.* Consejo de Cuenca Lerma-Chapala, Guadalajara.

de Anda, J., Quiñones-Cisneros, S.E., French, R.H. and Guzmán, M. (1998) Hydrologic balance of Lake Chapala (Mexico). *Journal of the American Water Resources Association* 34, 1319–1331.

de P. Sandoval, F. (1994) *Pasado y futuro del Lago Chapala.* Gobierno de Jalisco, Guadalajara.

Flores-López, F.J. and Scott, C.A. (2000) *Superficie agrícola estimada mediante análisis de imágenes de satélite en Guanajuato, México.* IWMI, Serie Latinoamericana No. 15. IWMI, Mexico City.

González-Villareal, F. and Garduño, H. (1994) Water resources planning and management in Mexico. *Water Resources Development* 10, 239–255.

Gorriz, C.M., Subramanian, A. and Simas, J. (1995) *Irrigation Management Transfer in Mexico. Process and Progress.* World Bank Technical Paper No. 292. World Bank, Washington, DC.

Guerrero-Reynoso, V. (1998) Participación social en el aprovechamiento sustentable de las aguas subterráneas – el caso de Guanajuato. In: *Memoria del Simposio Internacional de Aguas Subterráneas,* 7–9 December. León, Guanajuato, Mexico, pp. 33–42.

Guerrero-Reynoso, V. (2000) Towards a new water management practice: experiences and proposals from Guanajuato State for a participatory and decentralized water management structure in Mexico. *Water Resources Development* 16, 571–588.

Johnson III, S.H. (1997) *Irrigation Management Transfer in Mexico: a Strategy to Achieve Irrigation District Sustainability.* IIMI Research Report 16. IWMI, Colombo, Sri Lanka.

Kloezen, W. (1998) Water markets between Mexican water user associations. *Water Policy* 1, 437–455.

Kloezen, W. (2000) Viabilidad de los arreglos institucionales para el riego después de la transferencia del manejo en el distrito de riego Alto Río Lerma, México. IWMI Serie Latinoamericana No. 13. IWMI, Mexico City.

Kloezen, W.H., Garcés-Restrepo, C. and Johnson III, S.H. (1997) *Impact Assessment of Irrigation Management Transfer in the Alto Rio Lerma Irrigation District, Mexico.* IIMI Research Report 15. International Irrigation Management Institute (IIMI), Colombo, Sri Lanka.

Kloezen, W. and Garcés-Restrepo, C. (1998) *Assessing Irrigation Performance with Comparative Indicators. The Case of the Alto Rio Lerma Irrigation District, Mexico.* IWMI Research Report 22. IWMI, Colombo, Sri Lanka.

Mestre, E. (1997) Integrated approach to river basin management: Lerma-Chapala case study – attributions and experiences in water management in Mexico. *Water International* 22, 140–152.

Monsalvo, G. (1999) *Sostenibilidad institucional de las asociaciones de riego en México.* IWMI, Serie Latinoamericana 9. IWMI, Mexico City.

Palacios, E. and Martínez, R. (1999) Aspectos constitucionales, legales e institucionales. In: *Politicas opcionales para el manejo de la sobrexplotación de acuiferos en México. Estudio sectoral.* World Bank/CAN, Mexico City, pp. 6.1–6.41.

Rap, E., Wester, P. and Pérez-Prado, L.N. (2004) The politics of creating commitment: Irrigation reforms and the reconstitution of the hydraulic bureaucracy in Mexico. In: Mollinga, P.P. and Bolding, A. (eds) *The Politics of Irrigation Reform.* Ashgate Publishers, Aldershot, pp. 57–94.

Scott, C.A., Zarazúa, J.A. and Levine, G. (2000) *Urban-wastewater Reuse for Crop Production in the Sater-short Guanajuato River Basin, Mexico.* IWMI Research Report 41. IWMI, Colombo, Sri Lanka.

Scott, C.A. and Garcés-Restrepo, C. (2001) Conjunctive management of surface water and groundwater in the Middle Río Lerma Basin, Mexico. In: Biswas, A.K. and Tortajada, C. (eds) *Integrated River Basin Management: The Latin American Experience.* Oxford University Press, New Delhi.

Scott, C.A., Silva-Ochoa, P., Florencio-Cruz, V. and Wester, P. (2001) Competition for water in the Lerma-Chapala Basin. Economic and policy implications of water transfers from agricultural to urban uses. In: Hansen, A. and van Afferden, M. (eds) *The Lerma–Chapala Watershed: Evaluation and Management.* Kluwer Academic, Dordrecht, pp. 291–323.

Silva-Ochoa, P. (ed.) (2000) *Unidades de Riego: La otra mitad del sector agrícola bajo riego.* IWMI, Serie Latinoamericana No. 19. IWMI, Mexico City.

Tortajada, C. (1999) Legal and regulatory regime for water management in Mexico and its possible use in other Latin American countries. *Water International* 24, 316–322.

Vázquez, M.L. (1999) Decretos de veda en el estado de Guanajuato. *Aqua Forum* 4, 20–21.

Wester, P., Marañón-Pimentel, B. and Scott, C.A. (1999) Institutional responses to groundwater depletion: The aquifer management councils in the State of Guanajuato, Mexico. Paper presented at the International Symposium on Integrated Water Management in Agriculture, 16–18 June, Gómez Palacio, Mexico.

9 Water Resources Planning and Management in the Olifants Basin of South Africa: Past, Present and Future

M. de Lange, D.J. Merrey, H. Levite and M. Svendsen

9.1 Introduction

The singular history of South Africa has created huge development challenges, along with important opportunities to address them. During apartheid, rights to productive resources such as land, water and minerals were controlled by the dominant minority, while the remainder of the population was concentrated into isolated 'bantustans' or 'homelands'. Although formal barriers to residence and economic activity were quickly dismantled in the wake of democratic elections in 1994, the legacy of earlier policies persists in rural areas. This manifests itself in huge income disparities; dense concentrations of black South Africans in the regions of the former 'homelands'; a skewed distribution of basic infrastructure such as roads, electricity and piped water supplies; extremely high levels of unemployment; and a dualistic agricultural economy.

In irrigation, this dualism is reflected in the disparity between the commercial and the subsistence or small-scale irrigation sector. Only about 100,000 ha of the national total of 1.3 million ha is currently worked by small-scale farmers. A sophisticated irrigation industry with extensive manufacturing and equipment distribution capacity serves the commercial sector, while small-scale farmers remain isolated from this capacity

by insecure land tenure, and lack of formal education, capital, market access and management experience. At the same time, the presence of such a sophisticated industry probably inhibits the development of a more informal irrigation services sector. One of the results of these disparities is that, in the new government, there is an almost exclusive policy focus on the historically disadvantaged sector, in contrast to the previous predominant focus on the commercial sector. The historically disadvantaged sector is receiving a spectrum of support to both deracialize commercial agriculture and create sustainable livelihoods for the 12 million people residing in the former bantustans. Although agriculture directly generates less than 5% of South Africa's gross domestic product (GDP), it is only now gaining recognition for its importance in combating widespread rural poverty and as a stabilizing factor in the national economy. Taking a broader perspective on the contribution of agriculture to GDP and including associated support services and agro-industries, agriculture actually accounts for more than 14% of the total. The GDP multiplier of agriculture is 1.51 overall. Further, out of an economically active population of 13.8 million people, at least 35% is directly or indirectly dependent on agriculture. About 10% of total export earnings of the country is from agriculture. Irrigation produces a quarter of the agricultural output on 11% of the cultivated land (Hirschowitz, 2000; Mullins, 2002).

One of the early actions of the post-apartheid government was to formulate a new and progressive water policy that mandated, among other things, integrated management of water resources at the basin level. The vehicle for basin level management is the Catchment Management Agency (CMA), which is intended to be the primary policy-making and management entity at the basin level. The country is presently engaged in implementing this policy, and in the process, confronting a number of very challenging issues. These include the task of developing integrated representative governance of the CMA in a bipolar social and economic environment, sharing costs among water-using sectors, and formalizing and reallocating water use entitlements in a context of growing water scarcity.

The present study analyses management of a South African river basin under the old regime and policies, and that envisioned under the new set-up. The analysis employs a matrix of essential functions and key actors to identify the extent of coverage, gaps and overlaps, extent of participation, and needs for coordination. It focuses on one of the pilot basins in which the new approach is being developed and tested – the Olifants.

9.1.1 Overview

9.1.2 The Olifants basin

The Olifants River basin is located in the northeastern corner of South Africa and southern Mozambique (Fig. 9.1).

The bulk of the South African part of the basin lies in Limpopo and Mpumalanga Provinces, with a small portion in Gauteng Province. The river flows from southwest to northeast and, upon leaving South Africa, enters Mozambique and joins the Limpopo river before discharging into the Indian Ocean about 200 km north of the capital Maputo.

The Olifants River is about 770 km long and, with its tributaries, drains 73,534 km². Under the new National Water Act, the CMA excludes the two northernmost tributaries, bringing the total drainage area under the future CMA to about 54,000 km², an area the size of Slovakia or Croatia.

The climate is semiarid, with rain falling primarily during the summer (November to March). Precipitation averages 630 mm and potential evaporation is 1700 mm.

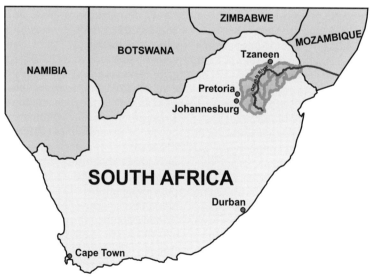

Fig. 9.1. Location of the basin in the region.

Although situated only 24° south of the equator, much of the basin is located at relatively high elevations (300–2300 m above sea level). This explains its cool winters and the wide annual temperature variation, which ranges from –4 to 45°C.

The population of the basin was 3.4 million in 2000. Population densities vary considerably, depending on whether a particular area was a former black 'homeland' (*bantustan*) or a part of the former white area, with densities ranging from 100 to 350 people/km² in former black areas and from 50 to 100 people/km² in former white areas. Whites currently comprise about 7% of the basin population. Ninety per cent of the black population lives in rural areas in the basin. The rate of illiteracy in the basin is more than 50%.

The basin is divided into five hydrologic areas generally regrouped in four ecological regions (Fig. 9.2).

- The Upper Olifants which correspond to the Highveld region;

- The Upper Middle and Lower Middle Olifants represent the Middleveld region;
- The Steelpoort basin assimilated to the Mountain area;
- The Lower Olifants situated in the Lowveld region.

9.1.2.1 Highveld region

The upper basin or Highveld region has a slightly higher rainfall than the Middleveld, and is characterized by extensive rainfed cropping and stock farming, coal mining and coal-fired power generation. Although the pollution impact of these activities is significant, the strategic importance is clear: 55% of South Africa's electricity is produced here. Annually, around 200 million m³ of water is imported, mainly from the neighbouring Vaal River basin, for water-cooling the power plants.

Another interesting characteristic of the upper basin is that the landscape is dotted with natural pans and farm dams. Users

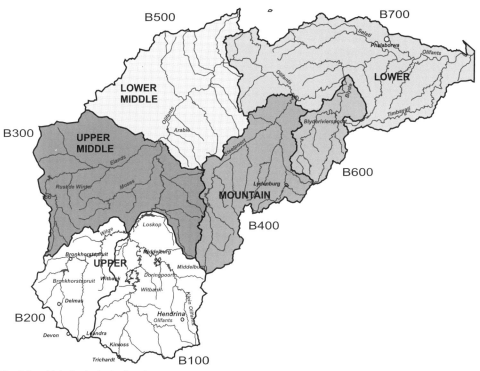

Fig. 9.2. Main hydrological regions.

in the Middleveld point out that they experience prolonged effects of droughts as a result, since these pans and dams in the upper basin have to fill up before significant runoff reaches the Middleveld.

9.1.2.2 Middleveld region

The Middle Olifants stretches for about 300 km from below Loskop dam to the dramatic drop down the escarpment near Hoedspruit. This escarpment area is famous for its natural beauty and exciting river rafting when the river flows full. However, this beauty and pleasure is in stark contrast to the plight of the 1 million people living in the old 'homeland' area in the Middleveld. This is one of the poorest regions in the country, where 75% of the population cannot read or write, and infant mortality, hunger and crime are rife.

The area directly below Loskop dam is quite different from the barren situation in the old homeland. Loskop dam was built by poor white labour as a public works programme after the Second World War. Below the dam are 28,800 ha of intensively irrigated land growing a variety of crops, with a trend towards permanent high-value crops including large citrus plantations and export table grape production under hail netting.

9.1.2.3 Mountain region

There is intensive irrigation in Steelpoort sub-basin and extensive mining activity in the valley. Expansion of platinum mining near the Steelpoort–Olifants confluence may result in the construction of another major dam on the Olifants River. This region is a heavy contributor to the basin water supply because of its high rainfall and steep slopes.

9.1.2.4. Lowveld region

Except for the upper part of this area (Blyde river sub-basin), there is little irrigation along the Olifants River below the escarpment, possibly because of poor soils. This region is characterized by game farms and industrial activity concentrated at the town of Phalaborwa, on the border of the famous Kruger National Park conservation area. The minimum dry season flow and the impact of the industrial effluent on the quality of the water entering the Kruger Park are of major concern to Park officials and conservationists.

9.2 Basin Hydrology

9.2.1 Supply

Basin water resources either originate from precipitation, which is transformed into runoff, evaporates or flows in the ground, or are imported from neighbouring basins.

9.2.1.1 Rainfall

Rainfall has an average value of 630 mm/year in the basin, and does not vary greatly at the level of the water management region (Table 9.1). At a smaller scale, however, the range is significant, varying from 500 to 1000 mm with values of up to 2000 mm along the Drakensberg escarpment.

There is a distinct rainy season between October to April, with the heaviest rain generally occurring in December and January, but the main characteristic of the precipitation is its high degree of unevenness, spatially, throughout the year and inter-annually. Midsummer dry spells are a common occurrence, often in January when dryland crops are at their most vulnerable.

Table 9.1.   Precipitation by region.

Water management region	Mean annual precipitation (mm)
Upper Olifants River	682
Upper Middle	621
Mountain	679
Lower Middle	550
Lower	631

Source: BKS (2002, Appendix C, p. 4-2 [Table 11]).

9.2.1.2 Surface water

The mean annual runoff (MAR) calculated under 'virgin conditions', that is, without the introduction of exotic vegetation (such as afforestation with eucalyptus species) and without the impact of human activities, would be 1992 million cubic metres (MCM) (Table 9.2). The average value of observed runoff with these influences present is around 1235 MCM at the mouth of the South African part of the basin, on the border with Mozambique.

A recent study of surface water resources (Midgley et al., 1994) presents relationships between rainfall and runoff by location in the basin. To generate significant runoff, rainfall has to be sizeable. In this regard, the mountainous regions play the most significant role. The Mountain region (Steelpoort sub-basin) as well as Blyde River (part of the Lowveld) provide about 42% of total runoff. It appears also that groundwater is of major importance in maintaining base flow in the river. The wide range observed in MAR follows the high inter-annual variability of the precipitation.

Typically, high flows occur from December to February and then decline until September. At the end of this period, the Olifants River can experience periods of no flow in the Kruger Park on the border with Mozambique. A major feature of the basin is also its capacity to generate extreme flows, yielding dreadful floods especially affecting Mozambique. During the last flooding period in February 2000, the flow in the Olifants peaked at 3800 m³/s at the mouth.

9.2.1.3 Groundwater

Total groundwater recharge in the basin is estimated by the Department of Water Affairs and Forestry (DWAF) to be 3–6% of the mean annual precipitation. This would amount to approximately 1800 MCM. Most of this recharge occurs during periods of heavy precipitation and it is suspected that the majority of water reaching the water table does so via macro pores (cracks, fissures, etc.) in the soil rather than through the actual soil body (Ashton, 2000).

Groundwater is an important source of water supply for many small towns, villages and small-scale farms, where it is used for stock watering and some irrigation, especially on the Springbok Flats in the Middleveld region. The largest share of the catchment's exploitable groundwater exists in a relatively shallow weathered aquifer that gives average yields in the vicinity of 1 l/s. Areas of higher potential (5 l/s and more) do occur in the area of the Steelpoort River, while roughly half of the catchment to the west of the Drakensberg Mountains may be classified as having moderate potential (1–3 l/s).

9.2.1.4 Storage, imports, exports and return flows

There are approximately 2500 dams in the basin, including 31 major ones defined as those storing more than 2 MCM. The total storage of major dams is 1100 MCM with a firm yield of 645 MCM per year. Small and minor dams supply additional storage capacity of 193 MCM. This storage capacity also represents an important source of loss by evaporation, estimated for all dams to be around 159 MCM.

Figure 9.3 shows the historical pattern of water resource development in the basin. As seen, periods of rapid growth took place in the late 1930s and again between 1970 and 1990. Plans for several additional dams in the basin exist, but await approval and funding.

Table 9.2. Runoff per sub-region.

Water management region	Mean annual runoff (MCM)	Range of mean annual runoff (MCM)
Upper Olifants River	466	134–1233
Upper Middle	200	86–538
Mountain	397	147–769
Lower Middle	107	23–555
Lower	822	255–2351
Total	1992	

Source: BKS (2002, Appendix C, p. 4-2 [Table 11]).

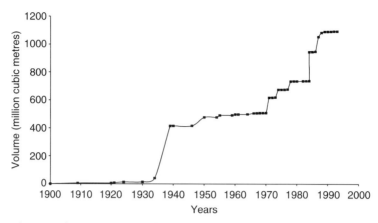

Fig. 9.3. Development of water resources in the Olifants basin (data from Midgley *et al.,* 1994).

Imports of water occur in the upper part of the basin, primarily for coal-fired power stations. Imports total around 200 MCM annually, and are derived from the adjacent Komati, Usutu and Vaal River basins. Comparable volumes of water are not reliably available in the upper reaches of the Olifants basin itself and the quality of the available water is often unsuitable. Small amounts of water are exported from the basin to the Limpopo provincial capital, Polokwane (5 MCM), and these exports are slated to expand in the future due to growing demand for domestic water supply there.

Return flows are not well estimated. BKS (2002, appendix C, Table 10) gives the figure of 42.3 MCM, mainly generated by urban areas in the Upper and Lower part of the basin. Return flows from mining activities are relatively small, varying from 0% to 20%, due to recycling processes. However, the high pollution levels in the river indicate that return flows that do occur are often heavily polluted. Little is known about return flows from irrigation, though they are undoubtedly significant, and this portion of the basin water budget requires further study.

9.2.2 Demand

There are various estimates for total water demand, even in the same report, suggesting the estimates are approximate at best.

BKS (2002, Table 3.2) provides total water demand figures of 1125.1 MCM/year in 1995 and a prediction it will rise to 1356.5 MCM by 2010. These figures include use of imported water and can be broken down among users as shown in Table 9.3. It should be noted the ecological and basic needs reserve, and the obligation to Mozambique as the downstream riparian, are not included.

9.2.3 Water balance

The basin is a water-stressed catchment and imports high-quality water from neighbouring catchments for economically important uses in the upper basin.

Nevertheless, in the Upper region there is a net surplus of 218 MCM per year currently, which is expected to drop to an estimated 168 MCM by 2010. In the Upper Middle, there is already a deficit of 60 MCM in 1995, growing to 79 MCM/year by 2010. The Lower Middle region shows a small surplus. Taking into consideration the need to meet ecological and downstream requirements, it is expected that water resources will be fully utilized by 2010 in the Upper, Upper Middle and Lower Middle regions.

For the Mountain region, the water balance shows surplus water available for further development. The Lower region appears to be in the same situation, with

Table 9.3.   Trends in water demand in the Olifants basin.

Use	1995	2010	Remarks
Irrigation	540.3	593.0	107,000 ha, 95% white commercial farming 9% increase projected mostly in the Upper Middle, Lower Middle and Lowveld regions Major crops: cotton, grapes, citrus, vegetables and grain
Livestock	27.8	33.1	Total of 2.3 million large stock units by 2010. Mainly in the Highveld and watered from boreholes (70%)
Urban (including smaller industries)	117.8	221.8	A 55% increase is likely to result from improving living conditions and population expansion, especially in the former homelands
Mining	94.3	93.8	In the Highveld region, open cast coal mining uses both water from the catchment and imports from the Komati, Usutu and Vaal Rivers. In the Mountain region, mines are a major user of both surface and borehole water. The Lowveld part of the basin contains numerous mines, but most of these are gold mines and have low water requirements, though they may have appreciable pollution potential
Afforestation	55.4	63.6	Indigenous forests will probably remain at 11,500 ha and permits for 15,000 ha of additional commercial afforestation will have been granted. It is envisaged that the existing 60,000 ha of exotic afforestation will only increase by a further 7000 ha
Reserve	None	460(?)	Determination underway but first figure of 460 MCM for the ecological Reserve is proposed in the Proposed National Water Resource Strategy (PNWRS, August 2002)
International requirements	?	?	Not determined. The Olifants is the largest contributor of the flow to Massingir Dam in Mozambique

Source: BKS (2002, Appendix C, Figs 2–7).

surplus water flows from the Blyde River sub-basin and the escarpment. The Selati River near Phalaborwa is experiencing water shortages, and transfers are planned from the neighbouring Letaba River basin to solve the problem (BKS, 2002, pp. 6-1–6-2).

Thus over the medium term, water in the basin appears to be adequate in terms of quantity, but there are serious distributional issues, related to both regions and user groups, that need to be addressed. Solutions contemplated comprise both construction of new storage and transfer facilities and real-location of available supplies among uses.

9.2.4 Water quality

The basin faces significant water quality problems, due to mining activities, indus-tries, power generation and agricultural use of water. The impacts of pollution from these activities (high salinity, high concentrations of metals, low pH) are probably multiple with serious ecological impacts. Of particular concern is the down-stream Kruger National Park, which is a major tourist attraction, as well as very worrisome health impacts, since many people rely on untreated surface water for drinking.

9.3 Legal, Policy and Institutional Environment

9.3.1 Water policy and water law

9.3.1.1 Pre-1994

With the establishment of a Dutch settle-ment at the Cape of Good Hope in 1652, customary principles governing access to water for stock watering and domestic use were supplemented and gradually replaced with ideas of European origin as the Cape Colony expanded. Over time, a pastiche of principles with roots in Roman, Dutch and English water law emerged to govern water allocation in South Africa (Thompson *et al.*, 2001). Initially, allocation policy employed the public trust doctrine, which gave the State (or the Dutch East India Company) the right to control and allocate use. Later a strong riparian component was added, giving individual landowners adjacent to natural watercourses the right to use water on those lands, subject to the rights of other similar landowners. Generally, the riparian principle was employed by white commercial large-scale farmers to secure water for irrigation. Appropriative rights to abstract water and use it elsewhere were also granted on a case-by-case basis. Water was classed as private or public, with water arising on a landowner's property entirely his to use. Flows in public streams were apportioned into 'normal' and 'surplus' flows, with different rules applying to each. These rules developed as the need arose from intensifying economic activity to regulate use among the commercial farms, mines and urban concentrations of the minority European population. In sum, during the apartheid era the white government, commercial farmers, mining firms and other interests established a complex system of formalized laws and institutions on a largely *ad hoc* basis, which excluded a large majority of the population (van Koppen *et al.*, 2003).

9.3.1.2 Post-1994

The introduction of democratic government in 1994 demanded changes in the skewed distribution of access to water for both basic and productive purposes. It offered a rare opportunity for comprehensive review of all water-related legislation to develop a modern water policy appropriate to a water-scarce region. This was done in a very deliberate and conscientious way under the farsighted Minister of Water Affairs and Forestry, Professor Kader Asmal (Box 9.1), and resulted in the new National Water Act of 1998 that is widely regarded as a model piece of legislation.

Two fundamental provisions in the 1996 constitutional bill of rights give 'basic

Box 9.1.	Water policy timeline.
May 1994	Review of all water related legislation initiated
Nov 1994	*Water Supply* white paper published
Nov 1996	*Fundamental Principles* paper approved by cabinet
April 1997	*Water Policy* white paper published
Sept 1997	Water Services Act promulgated
Aug 1998	National Water Act promulgated

rights' status to access to water to support life and for personal hygiene and a healthy environment, both now and for future generations. These constitutional rights undergird two fundamental pieces of new water legislation, the National Water Act (NWA) and the Water Services Act (WSA). The new water policy, as embodied in the White Paper on National Water Policy for South Africa, marks a radical departure from the previous legal regime. Basic principles are shown in Box 9.2.

South Africa's new water law reserves committed outflows to neighbouring countries. South Africa is a signatory to a SADC *Protocol on Shared Watercourse Systems* based on the Helsinki Rules, the Dublin principles and Agenda 21 of the Rio Earth Summit. The SADC protocol strives to maintain a balance between developmental needs of the member countries and the need for environmental protection and conservation. Signatories to the protocol commit themselves to seeking peaceful solutions to disputes. The protocol provides for the formation of basin-wide commissions, such as those for the Okavango and the Limpopo.

The new water law principles declare all water in the hydrological cycle to be an indivisible national asset, held in trust by national government (Box 9.2). Highly fragmented resource management across apartheid borders is being replaced by a catchment-based management system that follows natural, rather than internal political boundaries. Privatization of services provision is possible, but strictly governed by national legislation, administered through new local government structures.

Box 9.2. Principles of the National Water Policy.

- All water in the water cycle is a part of an indivisible national asset
- This asset is held in trust for society by the national government
- Water to meet basic human needs, to sustain the environment, and to meet legitimate needs of neighbouring countries is reserved
- All other water uses must be beneficial in the public interest
- The riparian system of allocation is abolished
- Allocations will no longer be permanent but for a reasonable period, e.g. 40 years, and can be traded
- Water resources will be managed on a catchment basis by specialized bodies
- All water use in the water cycle is subject to one or more charges intended to reflect the full financial costs of protecting and managing the water resource
- Water-based waste disposal is subject to appropriate charges
- Charges for water for basic human needs and for small-scale productive purposes may be waived for disadvantaged groups

Source: White Paper on a National Water Policy for South Africa (1997)

Since promulgation of the NWA, the water users have by-and-large remained the same, but there is a much greater emphasis on basic service provision to historically marginalized communities. The Government has embarked on a major investment programme to provide basic water supplies to the poor. New legislation also enables the establishment of water users' associations (WUAs) for user-management of shared local infrastructure – an option previously open to commercial farmers only. Furthermore, the NWA establishes mechanisms for public participation that enable reallocation of water to redress past racial and gender discrimination.

9.3.1 Actors

9.3.2.1 Government agencies

DWAF   During the apartheid era, the national DWAF was responsible for water resources management in the white areas of the Republic, while similar departments in the former homelands and self-governing territories concentrated largely on basic services provision. This fragmentation created severe problems in water resources management. In the white areas, water services were rendered by well-resourced municipalities while few resources were available to homeland water departments, leading to huge disparities in service levels between the white and homeland areas.

The 1996 constitution established water as a 'national competency', vesting responsibility for water resources and services in the DWAF. The character of the Department was fundamentally transformed in terms of its functions and staff profile (i.e. race, gender, discipline) to respond to its new mandate. DWAF embarked on an aggressive programme to speed up basic water and sanitation service delivery to the marginalized areas and changed the resources management paradigm from a supply-driven to demand-driven approach.

The NWA requires DWAF to develop a National Water Resources Strategy (NWRS), to create CMAs in the 19 Water Management Areas and to delegate catchment-based water resources management functions to these CMAs. Each CMA is to develop its own catchment management strategy through a participative process with water users in its area.

Through the WSA, DWAF designates local government structures as water services authorities (WSAs) and local government, water boards or private sector entities as Water Services Providers (WSPs).

NATIONAL DEPARTMENT OF AGRICULTURE AND PROVINCIAL DEPARTMENTS OF AGRICULTURE After 1994, nine provinces were created integrating the former four white provinces, six 'self-governing' territories and four 'independent' states. In the 1996 constitution, agriculture is a 'concurrent function', meaning it is a shared responsibility between the new National Department of Agriculture (NDA) and the nine Provincial Departments of Agriculture (PDAs).

PDAs still retain significant responsibility for irrigation infrastructure created previously. In the Limpopo Province, for example, the PDA has inherited over a hundred irrigation schemes of various types. Many are no longer functioning because the subsidies previously provided have been withdrawn and no new farmer-based structures are yet in place. The Province has launched a very ambitious 'revitalization' programme to organize farmers into WUAs and help them achieve the capacity to take over and manage their irrigation schemes profitably.

The former homeland departments of agriculture had played a significant role in creating irrigation infrastructure (often through parastatal corporations) and some of the PDAs have inherited responsibility for the operation, maintenance and refurbishment of these schemes. National policy currently promotes Irrigation Management Transfer (IMT) to the users, though progress in implementing this policy to date has been slow.

The homeland governments' role in water allocation for agriculture resulted from the link between land and water contained in the Water Act of 1956. Extension officers, together with tribal authorities, controlled the issuing and repeal of 'Permission to Occupy' (PTO) certificates to smallholder farmers, which conveyed rights to both land and water. Through the repeal of the riparian principle under the new NWA, the direct link between land and water allocation has been broken.

OTHER GOVERNMENT DEPARTMENTS   Before 1994, several departments played strong roles in water resources management, while more control now vests in DWAF through the new NWA. However, the Department of Mineral and Energy Affairs and the Department of Environmental Affairs and Tourism still play important roles in the protection and restoration of water resources through their requirements for Environmental Planning and Management Reports and for Environmental Impact Assessments, respectively.

FORMER HOMELAND GOVERNMENTS   In pursuit of the goal of separate institutions for each racial and ethnic group, apartheid-era governments attempted to divide the country gradually into various independent and autonomous racially homogeneous States. National States had a large degree of independence, while self-governing territories were to be gradually led to independent State status. Virtually no country other than South Africa recognized this system. The Olifants basin encompassed small portions of the national State of Bophuthatswana and included sections of three self-governing territories – Lebowa, KwaNdebele and Gazankulu (Thompson *et al.*, 2001). There were various departments in these homeland governments that dealt with water-related matters, but because only certain functions were transferred to the homeland administrations, others being retained by the Republic of South Africa, there was considerable confusion over some basic functions, such as the issuance of water use permits. Homelands were abolished in 1996 with the adoption of the national constitution.

SOUTH AFRICAN DEVELOPMENT TRUST AND OTHER PARASTATALS   The Development Trust, created originally in 1936 as the South African Native Trust, was responsible, among other things, for supplying basic infrastructure in the homelands, including water-related infrastructure. It did this directly and acting through the various homeland Development Corporations which it was responsible for establishing, financing and maintaining. It was dissolved with the implementation of the new constitution in 1996 and its water-related functions transferred to DWAF.

TRADITIONAL LEADERS   Historic tribal leadership in the Olifants basin was seriously undermined during the apartheid period. For instance, in an area that would traditionally be the domain of four tribal chiefs, the apartheid government ordained more than 100 new chiefs, mostly drawn from among lower ranking leaders who were willing to cooperate with the apartheid regime. This fragmentation, together with an influx of people through forced removals from white areas, created serious conflicts over tribal boundaries. These new 'traditional leaders' served in homeland government structures and areas of resistance to the regime were neglected in development initiatives. They had considerable authority over access to resources such as water and land (van Koppen *et al.*, 2003).

In the new set-up, tribal leadership is again at the heart of deep conflict, this time over shared local governance with new democratically elected structures. In the deep rural areas, members of the older generation, especially, swear allegiance to their chiefs and believe that the spiralling crime rate is at least in part related to the disrespect for the traditional systems and leadership. In turn, the new democratic structures are battling to establish themselves with limited resources and virtually no experience or systems for service delivery. In an attempt at integration, the tribal heads are automatically members of the new local councils, while other members are elected. As one community in the Olifants basin recently pointed out to their chief in a public meeting: 'these youngsters will come and go depending on election results, but you will stay, therefore we need you to take the lead for the sake of our development.'

LOCAL GOVERNMENT/MUNICIPALITIES   Local government is responsible for water services delivery, while the new CMAs will be responsible for water resources management. Local government structures, in turn, will be represented in the Olifants basin CMA.

Before 1994, 'municipalities' were established for service delivery in all cities and white towns of 200 houses or more. Their functions were limited to white urban concentrations, while rural areas were the responsibility of regional services councils. Separate structures were responsible for service delivery to blacks in the homelands.

After 1994, these boundaries were abolished and a period of restructuring followed. Most recently, a system has been adopted through which the entire country is divided into three types of local government structures: metropoles (class A) and shoulder-to-shoulder local municipalities (B), which are grouped together into 42 District

Municipalities (C). In effect, the former municipalities have seen their boundaries redrawn to include the rural hinterland.

WATER COURT/TRIBUNAL   Under the 1956 Water Act, a specialized Water Court ruled on disputes between rights holders. In the new legal framework, water use entitlements will be issued by the CMAs according to a water allocation plan that will form part of the catchment management strategy. People can appeal to the Minister on issues of unjust administrative process, followed by appeal to a newly established Water Tribunal.

9.3.2.2 Boards and agencies

CMAS   The NWA mandates the Minister of Water Affairs and Forestry to develop catchment-based water resources management through the creation of CMAs in the 19 Water Management Areas. Each CMA will develop and review, on a 5-yearly basis, its own catchment management strategy through a participative process with water users in its area. As the CMA develops the necessary capacity, DWAF will progressively delegate functions to it. This is not true devolution of power, but rather a delegation, since the Minister has the right to revoke authority if he is convinced that a CMA is falling short of its commitments.

The CMA board must be representative of the range of water users in the catchment. The Minister appoints board members through the following process. First (s)he appoints an advisory council to identify water user groups and relevant institutions in the basin. The Minister invites these institutions or organizations to nominate candidates for appointment to the CMA board. If (s)he is not satisfied with the suitability of a nominee, (s)he can ask for an alternative. The Minister then appoints the nominees and if necessary, 'tops up' the board with additional members to achieve racial and gender balance.

The CMA board appoints a Chief Executive Officer to develop and manage the CMA office or operational structure. This office may develop its own capacity to perform the full range of water resources functions, or it may contract out some or all of these functions. The NWA authorizes the CMA to collect a catchment management charge from water users to cover its costs.

As of early 2003, no CMA has yet been established, though proposals for several, including for the Olifants basin (BKS, 2002), are nearly ready to be submitted to the Minister. In the case of the Olifants, there has been a long period of attempting to consult with stakeholders in the basin as a basis for preparing a proposal. While the well-organized commercial sectors have participated actively, it has been relatively difficult to find effective means to consult the poor people scattered throughout the basin, particularly those residing in the former homeland areas (Wester *et al.*, 2003). Such consultation will be a continuous process, which will hopefully be done more effectively through the CMA when it is established. The proposed name of the CMA is the *Lepelle Catchment Management Agency*.

WATER BOARDS AND OTHER WSPS   Water boards were developed to provide services and perform water management functions beyond the reach of the old municipalities. Increasingly, they engaged in regional water supply schemes primarily for industrial and domestic use. Water boards are now regulated, together with other WSPs, under the WSA where their role in water resources management is considered incidental rather than intentional.

IRRIGATION BOARDS AND WUAS   The 1956 Water Act provided for the establishment of membership-based organizations called irrigation boards, through which groups of farmers could join forces to develop infrastructure and jointly manage their water supply – essentially a type of WUA. These irrigation boards were eligible for a one-third capital subsidy on shared water supply infrastructure, but membership was legally restricted to people who had title to the land receiving services from the irrigation board. This effectively excluded black membership in white irrigation boards, since blacks were excluded from land ownership in 'white

areas', but there was also no similar institution available to groups of black farmers with similar needs. Irrigation services in the homelands were supplied by government or parastatal corporations and participating farmers were passive recipients with no power to demand adequate services.

The NWA of 1998 calls for the transformation of existing irrigation boards into WUAs and removes title deed as a membership requirement. It thus also enables the establishment of WUAs on communally owned tribal or State land. Indeed, the NWA authorizes the issuance of water use entitlements – and by extension membership in WUAs – to water users rather than landowners. This is of particular importance in tribal areas where Permission To Occupy (PTO) certificates have traditionally been issued to men, but where women are predominantly the users of the land and water.

As of early 2003, only a few irrigation boards nationally have been officially transformed into WUAs, and only one formal WUA has been established on a small-scale irrigation scheme in the Olifants basin.

9.3.2.3 Farmers

COMMERCIAL IRRIGATORS   Irrigation farmers account for more than half of national water use and a quarter of the agricultural output in South Africa, irrigating in all 1.3 million ha. The commercial irrigation sector produces a wide range of crops for export and local use. It is supported by a sophisticated irrigation equipment manufacturing industry – indeed, micro irrigation technology originated in South Africa before it was developed into a major industry in Israel and the USA. Backward and forward linkages to input suppliers, service providers, processing, value-adding and export industries are not accurately quantified but are significant, both in terms of the national economy and employment.

Commercial irrigators fall roughly into three categories in terms of their access to water. About one-third of the irrigated area falls under irrigation boards or WUAs as discussed above, while another third is served from government water schemes, most of which are at some stage of handover to user management. The balance derived their water rights from the riparian principle and withdrew water directly from rivers and streams. The latter group had no need to participate in user management groups, a situation likely to change with the implementation of catchment-based water resources management.

Commercial irrigation in the Olifants River basin is almost a microcosm of the national situation, with farmers in government schemes and irrigation board schemes and unassociated farmers, as discussed above, growing a wide variety of crops. Basin farmers also represent a wide cross-section of attitudes to the new political reality, from the most conservative to the most liberal. Commercial farmers are well organized, but often overwhelmed and uncertain about the implications of the NWA. Many are also concerned about the affordability to water users of the establishment and operation of a CMA.

EMERGING IRRIGATORS AND OTHER VILLAGE-LEVEL ECONOMIC ACTORS   Some of the poorest rural areas in the country are found in the Olifants basin. Smallholder irrigation has had a troubled history of imposed development and State-managed irrigation infrastructure in a context of serious mistrust between the 'benefactor' and 'beneficiaries'. The 1600-ha Arabie–Olifants and other smallholder irrigation schemes in the basin have suffered in varying degree from the dependency created by this approach. Only a small portion of the irrigated area in the Olifants basin is occupied by smallholders, but a relatively large number of families derived at least part of their livelihoods from irrigation, either on the formal government schemes or on much smaller communal vegetable gardens or homestead food gardens. Not surprisingly, recent withdrawal of government management and subsidies for the schemes resulted in the collapse of most of the production here, with serious effects on food security in the area. However, amidst the collapse of the large systems, communal gardens and informal food producers have largely continued their activities as before.

In these poverty-stricken areas, the opportunities for expansion of both food production and income generation from agriculture and other small-scale village-based economic activities (such as brick making, ice making, poultry rearing, food processing) are largely dependent on the availability of water.

While their need for access to water is desperate, this sector is probably the most disorganized and under-represented of all water user sectors in the Olifants and many other South African river basins. The concept of small-scale water users forums (SWUFs) with representation in representative, decision-making and operational structures within the CMA framework has been proposed in the draft CMA proposal to address this problem (BKS, 2002).

9.3.2.4 Industrial firms

MINING COMPANIES   Coal mining dominates in the upper or Highveld region of the Olifants basin, while the mining and processing of platinum and related minerals is a growth sector in the Middleveld. Copper and phosphates mining and associated industries are the most important economic actors in the Lowveld area.

POWER GENERATOR (ESKOM)   The Olifants basin is of national strategic importance since more than half of the country's electricity is generated in coal-fired power stations in the coal-rich Highveld region of this basin. The electricity provider, ESKOM, depends on the transfer of high-quality water from other basins for its cooling towers.

FOREST PRODUCTS FIRMS   Two national forestry and forest products companies are active in the Olifants basin. Although the forestry activity in the Olifants is limited compared to other basins, the industry has been well represented in public consultations. Through the NWA, commercial forestry has been declared a Streamflow Reduction Activity (SFRA), which is viewed as a water use under the Act and therefore subject to all provisions in the Act that applies to other water uses, including allocation rules, water charges and measures for the protection of water resources.

TOURISM FIRMS   Several tourism companies operate river rafting and other recreational attractions, especially in the escarpment area. River rafting and fishing is entirely dependent on the flow in the river and therefore on dam operations, and high water quality is also important. Because of the small size of this sector, however, it is unlikely that tourism firms will significantly influence water resources management decisions Like many other users, though, these firms stand to benefit from improved and timely information on water releases from the major dams in the system.

9.3.2.5 Environmental advocacy groups

The Olifants is one of six significant rivers crossing the Kruger National Park. Pressure on mining companies and other industries because of their impact on Olifants water quality as it enters the Park, resulted in the founding of the Olifants River Forum (ORF) in the early 1990s. In the early stages of consultation to form a CMA, there was a vague expectation that the ORF might become the CMA, but it quickly became apparent that its history and limited membership base excluded this possibility. ORF now sees its role as that of an environmental watchdog and remains very active.

9.3.2.6 Development banks

Water infrastructure in the past has been funded almost exclusively by government. In the case of the former homelands, funds for irrigation development were administered through the Development Bank of Southern Africa. The new economic policy promotes private financing and requires transparency and sustainable financing, not only of infrastructure development, but also of its operation, maintenance and future replacement. It is therefore expected that commercial financing institutions will increasingly play a role in the construction and rehabilitation of infrastructure.

9.4 Essential Functions for River Basin Management

Chapter 1 identified 11 essential functions of basin management. A slightly modified listing of these functions is used in Tables 9.4 and 9.5, crossed with the key actors identified in the previous section. These functions are replicated, as appropriate, across four broad categories – surface water, groundwater, wastewater disposal and diffuse returns, or agricultural return flows. Cells are marked to indicate an actor's level of activity in a particular functional area. Information is drawn from interviews, printed materials and Internet postings. A number of interesting points emerge from a comparison of Table 9.4, which reflects the situation during apartheid (pre-1994) and Table 9.5, which conjectures about the possible scenario 5 years after the expected establishment of a CMA for the Olifants. This projection is made on the basis of a group discussion with key officials in DWAF in July 2001.

A few observations emerge from an examination of Tables 9.4 and 9.5.

- Prior to 1994, DWAF was the preeminent organization in managing surface water and wastewater, playing a major role in planning, allocating, supplying and protecting surface water resources.
- Roles of many key actors are changing, none so dramatically as that of DWAF. In combination with the CMAs, most of the same functions previously covered by DWAF will be covered in future. However, DWAF will delegate considerable authority to the CMAs and will assume new responsibilities in environmental protection.
- The important flood protection function, previously handled by DWAF and provincial and local government authorities, appears to have no primary locus of responsibility in the future. The CMA role is uncertain.
- Many uncertainties remain regarding who will do what in the future. In particular, the roles to be played by

Table 9.4. Past basin management functions and key actors – pre-1994.

Key actors	Surface water									Groundwater							Waste water						Diffuse returns				
	Plan (basin level)	Allocate water	Construct facilities	Distribute water	Maintain facilities	Monitor quality	Ensure quality	Protect against flooding	Protect ecology	Plan (basin-level)	Allocate water	Construct facilities	Withdraw/distribute water	Maintain facilities	Monitor quality	Ensure quality	Authorize discharges	Construct facilities	Operate facilities	Maintain facilities	Monitor quality	Enforce quality	Authorize discharges	Construct facilities	Maintain facilities	Monitor quality	Enforce quality
DWAF	●	●	●	●	●	●	●	●	○	○	○						●				●	●					
NDA/PDAs			●					●																			
Other government departments	○								●							●											
Water tribunal		●														○											
SA Development Trust and other parastatals			●	●	○																						
Homeland governments	○	○	●	●	○			○				●	●	○				○	○	○	○						
Municipalities			●	●	●			●				●	●	●				●	●	●	●						
Traditional leaders		○																									
CMA									○																		
Water boards			●	●	●																						
Water service providers																											
Irrigation boards			●	●	●																			○	○		
WUAs (umbrella)																											
WUAs (grassroots)																											

continued

Table 9.4. *Continued.*

Key actors	Surface water — Plan (basin level)	Allocate water	Construct facilities	Distribute water	Maintain facilities	Monitor quality	Ensure quality	Protect against flooding	Protect ecology	Groundwater — Plan (basin-level)	Allocate water	Construct facilities	Withdraw/distribute water	Maintain facilities	Monitor quality	Ensure quality	Waste water — Authorize discharges	Construct facilities	Operate facilities	Maintain facilities	Monitor quality	Enforce quality	Diffuse returns — Authorize discharges	Construct facilities	Maintain facilities	Monitor quality	Enforce quality
Emerging irrigators																											
Commercial irrigators			○		○							●		●										○	○		
Other small-scale economic actors																											
Non-irrigating farmers			○		○							○		○													
Mining firms			○		○							○		●				○	○	○	○						
Power generator (ESKOM)																		○	○	○	○						
Tourism firms																											
Forest products firms																											
Other manufacturers																											
Environmental advocacy groups									○																		
Development banks			●															●									

●, Indicates activity; ○, indicates limited activity.
DWAF, Department of Water Affairs and Forestry; NDA, National Department of Agriculture; PDAs, Provincial Departments of Agriculture; CMA, Catchment Management Agency; WUA, Water Users' Association.

Table 9.5. Projected basin management functions and key actors – 5 years after CMA establishment.

Key actors	Surface water									Groundwater							Waste water						Diffuse returns				
	Plan (basin level)	Allocate water	Construct facilities	Distribute water	Maintain facilities	Monitor quality	Ensure quality	Protect against flooding	Protect ecology	Plan (basin-level)	Allocate water	Construct facilities	Withdraw/distribute water	Maintain facilities	Monitor quality	Ensure quality	Authorize discharges	Construct facilities	Operate facilities	Maintain facilities	Monitor quality	Enforce quality	Authorize discharges	Construct facilities	Maintain facilities	Monitor quality	Enforce quality
DWAF	○	○	●	?	?	○	●	○	●						○	●											○
NDA/PDAs			○					○	○																		○
Other government departments									●							●											
Water tribunal		●							●																		
SA Development Trust and other parastatals							○				●					○											
Homeland governments						○																					
Municipalities			●	●	●							●	●	●	●			●	●	●	●						
Traditional leaders				?	?			?	?																		?
CMA	●	●	●	●	●	●	●		●	●	●				●	●	●				●	●					
Water boards			●	●	●	●						●	●	●	●												
Water service providers			●			○						●	●	●	●												
Irrigation boards															●												
WUAs (umbrella)		○	○	●	●						○																
WUAs (grassroots)			●	●	●						?	●	●	●													

continued

Table 9.5. *Continued.*

Key actors	Surface water									Groundwater							Waste water						Diffuse returns				
	Plan (basin level)	Allocate water	Construct facilities	Distribute water	Maintain facilities	Monitor quality	Ensure quality	Protect against flooding	Protect ecology	Plan (basin-level)	Allocate water	Construct facilities	Withdraw/distribute water	Maintain facilities	Monitor quality	Ensure quality	Authorize discharges	Construct facilities	Operate facilities	Maintain facilities	Monitor quality	Enforce quality	Authorize discharges	Construct facilities	Maintain facilities	Monitor quality	Enforce quality
Emerging irrigators																											
Commercial irrigators			○		○							●	●	●													
Other small-scale economic actors																											
Non-irrigating farmers			○		○							○	○	○													
Mining firms			○		○							○	●	●				●	●	●	●						
Power generator (ESKOM)																		●	●	●	●						
Tourism firms																											
Forest products firms																											
Other manufacturers																											
Environmental advocacy groups									●																		
Development banks			●															●									

●, Indicates activity; ○, indicates limited activity.

DWAF, Department of Water Affairs and Forestry; NDA, National Department of Agriculture; PDAs, Provincial Departments of Agriculture; CMA, Catchment Management Agency; WUA, Water Users' Association.

DWAF and the CMA in operating and maintaining facilities that deliver surface water are undefined.

- Some institutions from the previous regime have now disappeared (such as the South African Development Trust and homeland governments) and new ones are still in the process of formation. The first CMAs should be established in 2003; WUAs began being established in 2002 as DWAF worked out procedures and principles.
- Prior to 1994, planning was largely a technocratic and political process with little participation by other players. Planning under the CMA is expected to be much more inclusive and participatory.
- Prior to 1994, there was virtually no regulation of groundwater allocation and use, or of groundwater quality. In future, both the CMA and the water tribunal are expected to be involved in water allocation, while DWAF, the CMA and other government departments are expected to ensure quality.
- The roles envisioned for the CMA span a wide range, including planning and allocation, wastewater regulation, and monitoring and enforcement of water quality standards. It remains to be seen whether all of these functions, particularly the enforcement ones, are appropriate to such a pluralistic institution.
- Virtually no formal activity in regulating the quality of agricultural return flows is envisioned in the near future. This reflects both the existence of other higher priority concerns and the difficulty of addressing these issues.

The following sections provide additional observations based on the consultation with the stakeholders around the essential functions framework.

9.4.1 Water resources management

Before 1994, DWAF dominated the management and regulation of surface water resources and water quality, but with limited jurisdiction in the homeland areas. There was limited activity in the planning and allocation or control of groundwater and diffuse returns.

The intention is that the Minister of Water Affairs and Forestry will progressively delegate these functions to CMAs, as the CMAs achieve adequate capacity to assume such functions. DWAF will develop and maintain a national resource classification system against which the CMA will manage its resources in accordance with agreed resource quality objectives. In effect, DWAF will be transformed from a planning and implementation agency to one that does national level planning, provides technical support to CMAs and local water service agencies, and regulates all of these. The structure and personnel of DWAF are being modified to meet these new responsibilities.

CMAs are intended to be more participatory and sensitive vehicles for managing basin water resources. They incorporate a mix of public and private characteristics but are still a work in progress and the exact mix of characteristics that will emerge remains uncertain. On the one hand, CMAs are intended to be self-financing and semi-autonomous. On the other, they are expected to represent the broader public interest in protecting water resources and the natural environment. How compatible these functions will prove to be, and whether they will prove affordable for a self-financing CMA remains to be seen.

9.4.2 Water services provision – domestic and industrial use

Before 1994, DWAF played a dominant role in bulk water supply, but was generally not involved in service provision to end users. For domestic and industrial use, this was the domain of the homeland governments, water boards and urban municipalities.

Several changes in the South African government system affect the way in which water services will be provided in future. Most importantly, the new local government operates in newly demarcated shoulder-to-shoulder municipalities that erase the fragmentation along racial boundaries and consolidate jurisdiction over functions formerly held by the homeland structures, tribal authorities, urban-focused municipal areas and regional services councils. These new municipalities serve as the new WSAs under the WSA of 1997, and can act as WSPs themselves or contract this function out to the private sector, water boards or public sector WSPs.

9.4.3 Water services provision – agricultural use

The dominant water users in agriculture are the white commercial irrigation farmers, accounting for more than half of national water use. Before 1994, roughly one-third of the 1.3 million irrigated ha was serviced through irrigation boards. Irrigation boards acted as a type of WUA, providing services to its members, occasionally also supplying water to the municipality of the nearby rural town. However, only properties with full title deed were eligible for membership, which excluded blacks, who were prohibited from owning titled land in white areas.

All of this is now changing. Land and water rights have been separated by law, and there are some black water users that must be accommodated in the transformation of irrigation boards to WUAs. In fact, most applications for transforming boards to WUAs in 2002 were rejected by DWAF because they were not sufficiently inclusive (Faysse, 2003). Small-scale irrigation farmers are getting organized into WUAs, albeit not rapidly. In stressed river basins, the licensing process could very well lead to reductions in water allocations to commercial farmers, though this remains to be seen.

The tables reflect a new idea that has recently emerged with regard to WUAs: the concept of distinguishing between 'grassroots' and 'umbrella' WUAs. Initial interpretations of the NWA held that where there were both commercial and small-scale emerging farmers in one area, they must all be members of one WUA. However, based on experiences in the Olifants, it was realized that it may often be useful to enable smaller groups of water users sharing a sub-scheme to organize around their water infrastructure, and the various smaller WUAs could be federated into a larger WUA to address mutual problems.

9.5 Enabling Conditions

The essential functions and actors' roles depicted in Tables 9.4 and 9.5 provide a static view of responsibilities. Additional attributes of well-functioning basin governance systems relate to its dynamics. Chapter 1 suggested a list of attributes termed enabling conditions to analyse these. Here we discuss a slightly modified version of this list with respect to the Olifants basin (see also van Koppen *et al.*, 2003; Wester *et al.*, 2003).

9.5.1 Political attributes

South Africa's negotiated transition from white minority rule to democracy is legendary and is embodied in its 1996 Constitution. This early success firmly established negotiation as the preferred *modus operandi* and representativeness, legitimacy, equity and sustainability became requirements of the new political environment. The introduction of democratic governance gave South Africa's post-1994 government an unprecedented mandate for change. This mandate for major change brought uncertainty that dislodged vested interests sufficiently to enable fundamental policy review. In reviewing the water law, Professor Kader Asmal's strong political leadership resulted in this window of opportunity being seized to introduce a new system of inclusive representation and balanced power in water resources decision making.

The new system has to create mechanisms through which water users across the board can make themselves heard and understood, enabling gradual and systematic redress of racial and gender inequities, whilst ensuring a secure base for economic growth. This is a major challenge in policy and implementation. Lack of access – through both lack of rights and lack of infrastructure – is a priority issue of the rural poor in the Olifants basin. Their adequate representation is hard to achieve: most live in remote areas, excluded from participation through the cost of transportation and a lack of organization. Industrialists and commercial irrigation farmers, by contrast, are well organized and have larger financial resources, better enabling them to participate in consultation processes. The pre-eminent challenge to the CMA process is to find ways to ensure representative participation of major stakeholding groups and levelling the playing field to allow the various stakeholders to interact with equivalent knowledge, authority and legitimacy.

The environment is nominally protected through the strong provisions in the NWA for priority allocations to maintain the integrity of the resource base. However, the Environmental Affairs Ministry and the environmental non-governmental community may lack the organized political clout to ensure that such provisions are adequately implemented.

The proposed CMAs ignore internal political boundaries, playing to both the strengths and weaknesses of the new national political scheme. The South African situation is in contrast to countries like Turkey, where a long history of strong and stable local government provided ideal vehicles for the establishment of their new WUAs. In South Africa, local government structures were deeply problematic: well-resourced white municipalities and regional councils historically excluded black representation, while structures in the former homeland areas were weak and now have to grapple with major ideological differences between traditional tribal leadership and the new democratically elected representation. Apart from the advantages from a natural resource perspective of managing along natural boundaries, the thinking was that this deliberate disregard for political boundaries would enable the CMAs to continue their

business throughout successive political changes. Nevertheless, the Constitution provides for 'cooperative governance' where there are overlapping functions. The CMAs will have to find ways to engage effectively with the local governments, who both represent the citizens' interests and have responsibilities to provide water and sanitation services.

9.5.2 Informational attributes

It has been argued that, on the one hand, the consultation process in the Olifants basin for establishing the CMA was not fully effective in reaching the large majority of poor stakeholders; and on the other hand, the long consultation process is delaying action to address some of the serious issues and problems of the basin (Wester *et al.*, 2003). One of the lessons of the water law consultations was the importance of trusted information as a basis for consultation and negotiation. Good information is crucial to delineate areas of agreement and disagreement to structure and inform debate, but of little use if the source is doubted. Equally, good information becomes useful in negotiation and decision making only when it is accessible by all interested and affected parties. South Africa is well equipped to use the most modern techniques for data gathering, storage and knowledge creation, but faces a major challenge in presenting information in a meaningful way to the wide range of interests in the sector. Those most in need of water for basic and productive uses are poorly equipped to access and interpret information from the national systems.

Indeed, good information has the power to defuse unnecessary tensions, while misinformation and lack of information are powerful instigators of conflict. For example, during public consultation on South Africa's largest inter-basin transfer scheme, from the Orange River system to the economic heartland around Johannesburg through the Lesotho Highlands, irrigating farmers along the lower Orange were concerned that their future was in jeopardy. A rumour spread quickly that the water crisis in Johannesburg was due to the estimated 6 million illegal immigrants in and around the city. A quick back-of-the-envelope calculation put an end to the rumour: the basic water needs of 6 million people would only irrigate about 5000 ha, a fraction of the irrigation along the Orange and less than 0.5% of the irrigated area nationally. However, information is often more complex than in this example, and open to interpretation.

The language of consultation is a particularly important tool in processes of inclusion and exclusion. Preparatory consultations in the local languages are a powerful technique to prepare the rural poor to engage with other water user sectors. This problem is recognized but not yet fully addressed.

9.5.3 Legal authority

South Africa possesses a strong legal framework for water resources management. Water is viewed as an indivisible national asset, held in custody by the State. Consequently, significant powers vest in the Minister of Water Affairs and Forestry. Legal authority is not expected to be a constraining factor in the implementation of basin management. Similarly, the CMAs and WUAs are provided a strong legal base in the NWA. Indeed the NWA promotes the establishment of CMAs, at the initiative either of water users in the Water Management Area or of the Minister.

9.5.4 Resources

It is often said by South Africa's neighbours that they lack the resources to conduct public consultation processes as comprehensive as those characterizing South African policy making. The ongoing cost of consultation during implementation was the subject of exhaustive debate during the water law review. Compromise was reached by toning down the specifics around consultation requirements in the NWA. Still

there are important unresolved questions regarding the potential for financial autonomy of the CMAs and how they will be funded.

Water resources management functions will be delegated to CMAs in 19 Water Management Areas. Currently, some functions are performed at DWAF head-office in Pretoria, while others are the responsibility of its regional offices. Demand may well outstrip supply of human, financial, institutional and infrastructural resources to service such a large number of water management offices. Modern water resources management is dependent on a range of specialist skills that may be unaffordable and unattainable by all 19 CMAs in the short to medium term. This possibility has led to discussions over the possible sharing of specialist expertise across CMAs. In the new political framework, another challenge would be to find leadership with command of the technical, social and political skills required by the job. Water managers of the previous era were highly skilled technically, but were not obliged to possess the wider range of skills necessary in the new framework.

9.6 Conclusion

Speculation on the future of integrated water resources management in the Olifants basin needs to be informed by the current policy and legislative framework, as well as historic factors and issues of water scarcity, and emerging critical issues. Some key themes which will characterize the future are outlined below.

- *There is a single source of authority and power.* The Constitution and the NWA vest the custodianship of South Africa's water in the Minister of Water Affairs and Forestry. While this enables DWAF to keep tight control on the CMAs' adherence to national policy as expressed in the NWRS, it implies that the process is externally induced and driven. At the same time, the CMA may be tempted to inflate its empire with functions better addressed at either national or local levels, exacerbating problems of sustainable financing. The ministry will have to proceed carefully to empower the emerging CMAs, and to refrain from casual interventions once they are empowered, if they are to develop a sense of responsibility and local accountability. Another distinct advantage of this single source of authority is that water allocations, called 'water use entitlements' in the NWA, are well defined.

- *Poverty and poor representation of the neediest sectors.* It is still unclear to what extent the CMA will have a developmental agenda focused on addressing poverty, though the latest draft of the CMA proposal (BKS, 2002) does recognize this role. Schreiner *et al.* (2002) suggest two possible scenarios: (i) 'public participation' that is captured by the powerful, thus reinforcing marginalization of the disadvantaged, and (ii) a scenario that builds poverty eradication and achieving gender equity into the initial design of the CMA. Even if the second scenario prevails, however, ultimately a balance must be secured by the disadvantaged themselves developing the organization and political muscle to protect their rights and privileges. The right to water for 'basic human needs' is a well-protected right in both the Constitution and the NWA, but lack of access to water for domestic use is still a very serious problem in the former homelands in the Olifants. Development of domestic water services, which is the domain of local government, is of the utmost importance to rural residents and is largely independent of the CMA. However, in the realm of small-scale economic activity, CMA activities could affect tremendously the ability of the poor to improve their livelihoods by making available additional allocations of water to these sectors. Information flows need to be designed to reach all water user sectors, but especially those with limited access to modern media.

- *From administrative to public processes for water allocation.* Water allocation was historically an administrative process with little public involvement. The NWA introduces mechanisms for public interaction at a relatively localized level, through the development of an 'allocation plan' in the CMA's 5-yearly catchment management strategy. This implies a shift from user-to-official interaction to a much more direct negotiation among user groups. In a context of growing water scarcity, conflict management is likely to become increasingly important. This is a vitally important issue. Powerful interests, such as mining and large-scale commercial sectors, still have very strong voices, and of course have strong arguments for protecting their access to water: mining and agriculture in the Olifants basin are major earners of foreign exchange. On the other hand, there are concerns as to whether this *status quo* is best for the millions of poor people living in the basin, with little or no access to water for their own productive uses. Further, there are large uncertainties regarding the degree to which there will be unmet water demands in the future in this basin and the extent to which these can be met by better demand management (Levite *et al.*, 2002).

- *Growing water quality issues.* Intensification of economic activities like mining, industry and farming, as well as growing population densities will put increasing pressure on the quality of water resources. Innovations like the existing agreements on quality management through 'controlled releases' of polluted water from coal mines during high-flow periods will gain importance, but remediation will eventually need to enter the next, more costly phase of treatment. Water quality impacts of dense settlements and agricultural outflows will demand attention, especially in the light of national standards, requirements for the maintenance of an ecological reserve, and achievement of resource quality objectives.

In sum, South Africa is at an early stage of its long journey to effective integrated management of its water resources. While its legislation and policies are ahead of those of many countries, it faces special major challenges. Both nationally and in many of its basins, it faces potentially serious gaps between growing demand for water and the available resources. It needs to focus far more than it has on how to use the water productively while also reducing the current gross inequities and achieving the aspirations of its poor majority to benefit from this and other resources.

References

Ashton, J. (2000) *Conceptual Overview of the Olifants River Basin's Groundwater, South Africa.* African Water Issues Research Unit, University of Pretoria.

BKS (2002) Proposal for the establishment of a Catchment Management Agency for the Olifants water management area. Department of Water Affairs and Forestry. Draft, November 2002. Appendix A, B, C, D, E.

Department of Water Affairs and Forestry. Proposed National Water Resource Strategy, first edition, August (2002) (also website http://www.dwaf.gov.za/Documents/Policies/NWRS/Default.htm) and its Appendix 9 D4, concerning the Olifants river basin.

Faysse, N. (2003) Possible outcomes of smallholders' participation in water resource management institutions in South Africa.

Draft paper under review for publication. IWMI, Pretoria.

Hirschowitz, R. (2000) *Measuring Poverty in South Africa.* Statistics South Africa, Pretoria.

Levite, H., Sally, H. and Cour, J. (2002) Water demand management scenarios in a water-stressed basin in South Africa. Paper presented at the 3rd WARFSA/Waternet Symposium, Dar-es-Salaam, Tanzania, 30–31 October.

Midgley, D.C., Pitman, W.V. and Middleton, D. (1994) *Surface Water Resources of South Africa 1990.* Water Research Commission, Pretoria.

Mullins, D. (2002) *Calculation of Sectoral Multipliers.* Conningarth Economists, Pretoria.

Schreiner, B., van Koppen, B. and Khumbane, T. (2002) From bucket to basin: a new paradigm for water management, poverty eradication and gender equity. In: Turton, A. and Henwood, R. (eds) *Hydropolitics in the Developing World: a Southern African Perspective.* African Water Issues Research Unit (University of Pretoria), Pretoria.

Thompson, H., Stimie, C.M., Richters, E. and Perret, S. (2001) *Policies, Legislation and Organizations Related to Water in South Africa, with Special Reference to the Olifants Basin.* IWMI Working Paper 19 (South Africa Working Paper 7). IWMI, Colombo, Sri Lanka.

van Koppen, B., Jha, N. and Merrey, D.J. (2003) *Redressing Racial Inequities Through Water Law in South Africa: Revisiting Old Contradictions?* Comprehensive Assessment of Water Management in Agriculture Research Paper 3. IWMI, Colombo, Sri Lanka.

Wester, P., Merrey, D.J. and de Lange, M. (2003) Boundaries of consent: stakeholder representation in river basin management in Mexico and South Africa. *World Development* 31, 797–812.

10 Water Resource Management in the Dong Nai Basin: Current Allocation Processes and Perspectives for the Future

Mark Svendsen, Claudia Ringler and Nguyen Duy Son

10.1 Introduction

Vietnam is emerging slowly from an era of strong, but compartmentalized, central control of all-important functions to one where limited authority is being delegated and more effective linkages among sectors and ministries are being established. One sector in which this is taking place is the water resource sector.

This chapter examines the policy and institutional environment governing water allocation and other management functions in the Dong Nai River basin, together with the rules-in-use, which currently control allocation decisions. In doing this, it employs the interactional analysis matrix developed in Chapter 1 and the concept of stages of basin closure, developed in Chapter 2. The chapter will define the basin and overview basin hydrology and water uses. It will then assess the present level of stress on basin water resources and its implications. Following this, it will describe the legal framework for basin management, identify the important stakeholders in basin water management, and examine the processes by which allocation and water control decisions are currently made. A final section

discusses possible future changes and the implications for agriculture.

10.2 Basin Hydrology

10.2.1 Boundaries

Vietnam is formally divided into nine basins for water resource planning. One of these is the basin entitled *Dong Nai-Sai Gon and Surrounding Areas*. It covers about 15% of the total surface area of the country, not including the roughly 10% of the total Dong Nai watershed, which lies in neighbouring Cambodia. The main stem of the Dong Nai River extends for 628 km, all within Vietnam.

While based primarily on watershed boundaries of the Dong Nai and tributaries, this planning unit includes a number of short coastal rivers that flow out of the mountains and across a narrow plain directly to the South China Sea. In addition, the West Vam Co River adjacent to the Mekong Delta is not included in the planning basin, though it is a tributary to the Dong Nai River near its mouth.[1]

---

[1] A primary reason for this is that the West Vam Co receives substantial transfers from the Mekong River through a connecting channel.

While the basin defined in this way is appropriate as a planning unit, for management purposes the Dong Nai basin is more properly defined as the hydraulic unit draining into the South China Sea through the Dong Nai River. This definition would exclude the coastal rivers in Ninh Thuan, Binh Thuan and Ba Ria-Vung Tau Provinces and include the West Vam Co River in Long An Province. The rationale for this is that the management unit should cover a hydraulically integrated area where water-related decisions in one part of the unit have the potential to affect other parts of the unit. It should exclude unaffected and unrelated areas. The southeastward-flowing coastal rivers generally do not satisfy this criterion and should thus be excluded from the management unit.

Management decision making relating to these small external watersheds should rest with local stakeholders and not with the larger basin council. This has two complementary advantages. First, it avoids unnecessarily expanding the basin council and complicating its decision-making processes. Second, it allows control of the smaller coastal basins to remain with local stakeholders, rather than transferring influence and control to the larger multi-party body.

Complicating this picture, however, is an existing transfer from the upstream Dong Nai River to the Cai River in the coastal area. Additional diversions are planned. Clearly planning for future inter-basin diversions needs to include all of the affected areas, justifying the current planning boundaries. For management purposes, however, unless these inter-basin diversions are the dominant source of water for the receiving basin, it is more appropriate to manage transfers through bilateral interactions between managers of the two basins. If the transfers do comprise the majority of flow in the receiving basin, then it may be reasonable to include the whole of the receiving basin in the purview of the management body for the larger basin and give receiving

basin stakeholders representation in this body.

Defined as a management unit, as above, the Dong Nai basin covers about 40,200 km² and comprises four major branches. These are the Dong Nai main stem, the Be, the Saigon and the two branches of the Vam Co (Fig. 10.1). The basin covers all or nearly all of five provinces[2] and a part of five others.

10.2.2 Characteristics

The Dong Nai basin has several distinct hydrogeological regions, ranging from the lowland areas in the Vam Co Dong River system that are inundated from Mekong floods during the rainy season, to the Central Highland areas of up to 1600 m. The lower basin reaches are subject to tidal influences, particularly during the dry season, with substantial saltwater intrusion. Precipitation averages 2000 mm, ranging from 1200 mm in the lowlands to 2800 mm in the highlands. The basin exhibits marked seasonal variations in flow with 87% of total precipitation concentrated during the rainy-season months of April/May to October/November. In addition, there are modest temporal variations in flow, with low inflows of 28 billion (10^9) cubic metres (BCM) in 1998 compared to high inflows of 40 BCM in 1997.

10.2.3 Basin water balance

10.2.3.1 Supply

The natural yield of the basin, averaged over the 1978–1998 period, is around 36.26 BCM annually. Imports from the Mekong to the Vam Co tributary contribute an estimated additional 6.22 BCM per year, while power generation diversions to the Cai basin reduce the basin water availability by about 0.64 BCM.

The current project inventory for the basin includes additional storage of

---

[2]   Ho Chi Minh City is considered as a province here.

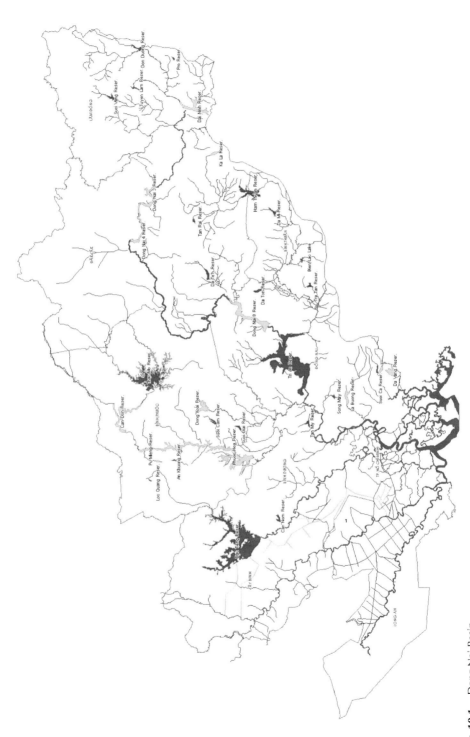

Fig. 10.1.   Dong Nai Basin.

4.23 BCM, with 1.95 BCM of that considered likely to be in service by 2010. By that year, basin storage would thus total about 7.56 BCM (Fig. 10.2). Still, the options available to system managers will remain limited as a consequence of the limited storage. Limited storage also means that seasonal variations in discharge and demand become extremely important, and that short-term water balances, rather than annual averages, will drive allocation and management decision making.

Significant development of storage began in 1963 with the construction of the Da Nhim dam and is continuing. Total storage in the basin is relatively small at 5.61 BCM, and comprises only about 13% of average annual discharge. All basins are different and have different characteristics and storage requirements for effective management. In general, however, storage is an essential element of water control and management and it is informative to contrast storage ratios in some other basins with that of the Dong Nai. Indonesia's Brantas basin, for example, lacks reservoir sites and can store only about 4% of annual discharge. On the other hand, Australia's Murray–Darling basin storage is about 250% of annual discharge, and on the Colorado River in the USA storage is about 400% of annual discharge. Other things being equal, this suggests that additional storage in the Dong Nai would enhance the scope of management options available to managers there.

10.2.3.2 Demand

Current demand for water in the basin is not known with certainty, but is estimated in Table 10.1. As can be seen, while domestic and industrial demands are important, irrigation and salinity control requirements are the dominant items. Estimates of the level of sustained dry-season reservoir releases required for salinity control in the lower river vary widely. Values used here are flows of 85 m^3/s at Binh An on the Dong Nai River from Tri An Reservoir and 25 m^3/s at the site of the future Ben Than intake on the Sai Gon River from Dau Tieng Reservoir. The aggregate salinity control requirement shown is intended to control the tidal salinity front at the points of withdrawal of the Ho Chi Minh City (HCMC) urban water supply, and does not directly consider any additional environmental or waste dilution needs.[3]

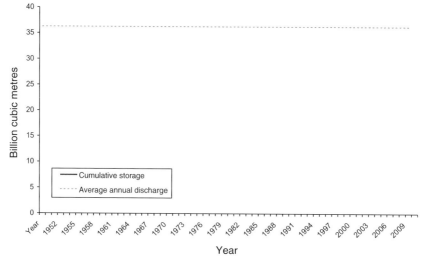

Fig. 10.2. Cumulative basin storage capacity, 1950–2010.

---

[3]   HCMC does receive some controlled deliveries for waste dilution.

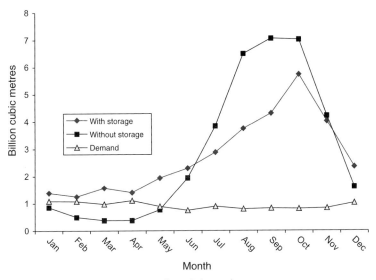

Fig. 10.3. Hydrographs of Dong Nai Basin supply and demand.

Table 10.1. Gross water demand in the Dong Nai basin.

Source	Current demand (BCM)
Irrigation	4.57
Domestic water supply	0.50
Industry	0.72
Salinity control	3.47
Hydropower	0.64
Total	9.90

One difficulty in estimating demand, implicitly for surface water, is the lack of current knowledge about ground and surface water interactions in the basin. This uncertainty affects both irrigation and domestic/industrial demands, but is most important for irrigated agriculture. While about one-fifth of domestic/industrial use is currently drawn from groundwater, future growth in this use sector is expected to be met almost entirely from surface water sources, making it reasonable to allocate domestic and industrial use to surface water demand.

Groundwater use for irrigation, however, is growing rapidly, and much of the new high-value crop production, such as coffee, fruit and pepper now being established relies on private pumping of groundwater. The extent to which groundwater

aquifers and surface water flows are interconnected in a given region is thus important in determining the share of demand that should be allocated to surface water resources. In Tay Ninh Province, for example, it is known that seepage from the Dau Tieng Reservoir and associated irrigation facilities have caused a rise in the regional water table from 10–12 m below the surface to just 4–5 m, facilitating private irrigation development and allowing extraction with cheaper centrifugal pumps. In other areas, private pumping is drawing down the water table. Shallow water tables are typically hydraulically connected with adjacent streams and rivers, and it is reasonable to attribute this private well demand to surface supplies until more detailed information is available which would permit partitioning into surface and groundwater supplies.

10.2.3.3 Balance

Viewed on an annual basin, there is still little stress in the basin, and seemingly little need to incur significant expense for coordination and management. Total current demand is only about 30% of annual supply, even assuming no reuse of municipal and industrial return flows.

The picture changes significantly, however, when supply and demand are examined on a monthly basis. Figure 10.3 shows monthly supply and demand averaged over 3 years, 1996–1998.

The *without storage* curve depicts the volume of synthesized inflows to streams and rivers throughout the basin. The *with storage* curve shows supply as reservoir releases from the three major storage reservoirs in the basin combined with natural inflows below each reservoir. The second curve thus shows the same volume of inflows as the first, but distributed over time according to the pattern of reservoir operations prevailing during the subject period. Existing storage is able to take the monsoon peak off the inflow hydrograph and provide nearly four times the dry-season flow in March and April than would have resulted without storage.

More important to observe, however, is that during the first 4 months of the year, supply and demand are nearly in balance, and with small increases in demand over time, or in poor rainfall years, the basin will be seasonally in deficit. This will place the domestic water supply for HCMC, located at the lower end of the basin, at risk, both in terms of absolute volume of available water and as a consequence of inadequate flows pushing the tidal salinity front below the HCMC water intake on the Dong Nai River at Binh An. This constraint can be relaxed in a number of ways, including shifting more hydropower production to the dry months when reservoir releases can do double duty. However, each alternative solution has its own set of benefits and costs.

The important point here is that, from a monthly and seasonal perspective, the basin is already approaching a stressed condition and pressure will be increasing to find the lowest-cost approaches that will relieve stress and ease constraints on the most affected sectors.

10.2.4 Water use sectors

10.2.4.1 Hydropower

Five reservoirs currently produce electricity in the basin (Table 10.2). The Dong Nai basin is one of two major hydropower producers in the country, the other being the Da River basin in the north. Installed capacity at the five Dong Nai hydropower stations is about 1.2 GW. The basin thus represents about 25% of the national hydroelectric capacity and about 18% of total national installed generating capacity of around 6.5 GW. It is thus an important power producer, but not the major one in the country. A 500-kV north–south inter-tie makes it possible to transfer power back and forth between the north and the south of the country in response to demand.

Staff at each of the power stations are employed by Electricity of Vietnam (EVN), a State-owned company established under the Ministry of Industry (MoI). Generation is controlled by an EVN unit, the National Regulation Center (NRC), in Hanoi. Under the NRC are two Regional Regulation Centers – one for the north, also in Hanoi, and one for the south, located in HCMC. Prior to the completion of the north–south inter-tie in 1994, Dong Nai basin power stations were under the control of the Southern Regional Center in HCMC. Since completion of the inter-tie, all power stations in the country are controlled from the national centre in Hanoi. The NRC prepares yearly, quarterly and monthly plans for power production for each power station in the country, thermal and hydro, and directs operations on a real-time basis. Power station personnel have little discretionary authority with respect to operations.

Table 10.2. Hydropower generation in the Dong Nai basin.

Dam	Design capacity (MW)	Annual output (GWh)	Active storage (MCM)
Tri An	400	1700	2542
Da Nhim	160	1025	156
Ham Thuan	300	957	523
Da Mi	175	595	17
Thac Mo	150	590	1260
	1185	4867	4498

Because thermal stations generally have higher operating costs, operators attempt to maximize the use of the hydro stations. Moreover, generation from hydro stations is generally increased during daily periods of peak demand to take advantage of their flexibility. Plants are thus operated for a combination of base load and peaking power. Daily and hourly changes in generating levels are ordered by the national centre through a computer network linking the stations and the centre. Data on actual releases and generation are also fed into this system by the power stations. These data are closely held by EVN.

In addition to generation targets, there is an operating curve for each reservoir giving maximum allowable reservoir elevations throughout the year. When a reservoir level exceeds that specified by the curve, spill is ordered. Spills are generally avoided whenever possible because they bypass turbines and result in a loss of potential energy generation. These curves are regarded as a part of the dam design, and are rarely changed. There is thus no dynamic allocation of flood storage based on long-range weather and climate patterns. Rule curves for Tri An Dam are shown in Fig. 10.4.[4]

10.2.4.2 Irrigated agriculture

The Dong Nai River basin boasts a highly dynamic agricultural sector, with products ranging from basic staples like rice and maize to raw materials for local industry including rubber, cashew (both non-irrigated) and sugarcane and to high-valued crops like coffee, fruit, grapes, pepper, tea and vegetables. It produces about 15% of the total national agricultural output (Ringler et al., 2001).

Low rainfall during the dry season (as little as 10–50 mm/month) and dry spells in the rainy season make irrigation indispensable for the cultivation of many crops. According to estimates by the Sub-Institute for Water Resources Planning (SIWRP), the gross irrigated area in the Dong Nai basin (excluding the area irrigated by the West Vam Co River) is 630,000 ha. The net area irrigated during the dry season is an estimated 293,086 ha or 24% of the total agricultural area.

Estimates of the share of groundwater irrigation in total irrigated area vary widely, ranging upward from a conservative 10–15% of the total to a third of all the crops in the basin. Groundwater-irrigated crops include the larger share of the coffee area, pepper and fruit trees, some upland crops,

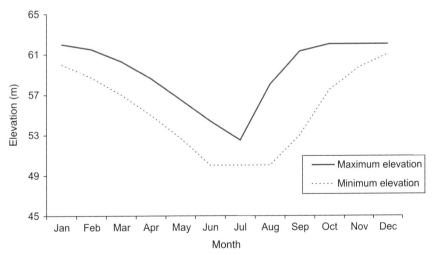

Fig. 10.4. Example of a rule curve for a basin reservoir.

---

[4]   Elevations shown for each month are for the first day of the month.

and even a small share of the area under rice. The higher estimate is supported by the rapid increase in the number of pumps in use in the basin. In 1999, pumps numbered an estimated 248,000, having grown at a rate of 16.1% per year during 1991–1999. This rate of increase is second only to the Mekong Delta, where private pumping from canals is prevalent.

Over the 1999–2010 period, Vietnamese planning authorities project growth in gross irrigated area in the basin of 3.85% per year, with a rapid increase in area under upland and industrial crops and perennial crops, and slowing growth in areas under rice (SIWRP, 2001). As there are no water quantity measurements in the basin's irrigation systems, irrigation water demand must be estimated. Estimates based on the FAO crop water demand methodology yield irrigation demands of about 4.6 BCM. Estimates based on a detailed farm production survey yield water demands of 3.8 BCM. The difference can be accounted, in part, by the large crop water demand of perennial crops indicated by the FAO methodology that, in effect, is being supplied from shallow groundwater due to the plant's deep root system instead of irrigation applications. Based on the SIWRP irrigation expansion planning for 2010, irrigation water demand will increase to 5.4 BCM by 2010, at a rate of 3.3%/year.[5]

10.2.4.3 Flood control

The Dong Nai basin is subject to flooding in both upstream and downstream areas. In downstream areas, flooding from rainfall received in the upper watershed can be exacerbated by flooding from the adjacent Mekong River, local rainfall and tidal influences. The official flood season is specified as 1 July to 30 November (MARD, 2000).

The three large upstream reservoirs in the system afford some flood control potential, but the level of protection provided depends on the way in which the reservoirs are operated. Presently, two of the three

largest reservoirs (Tri An and Thac Mo) are operated primarily for hydropower production, while the third (Dau Tieng) is not equipped for power production and is operated for irrigation, domestic and industrial water supply, and downstream salinity control. The operating regulations for Dau Tieng describe operation for flood control and require consideration of Tri An and Thac Mo operating plans. In all three cases, however, emphasis appears to be on operation to fulfil the reservoir's primary purpose – irrigation and hydropower generation, respectively. Even releases from Dau Tieng that are well below spillway design capacity can cause flooding and damage in downstream Binh Duong Province, where habitation and economic development is present on the flood plain below the dam.

The hazards of this single-purpose operating regime were demonstrated in October of 2000, when continuous heavy rains over much of the watershed resulted in the need to spill water simultaneously from all three reservoirs (Ngoc Anh, 2000). The rains occurred near the end of the rainy season when reservoirs were nearly full and spill was required to protect the dams. Major spills occurred for 10 days on 10–20 October. Downstream flooding in the basin resulted in 18 deaths, more than 37,000 dwellings damaged or destroyed, while affecting more than 60,000 ha of crops.

In addition to calling for construction of additional reservoirs and river dyking, a report on the 2000 flooding prepared by SIWRP suggests a number of new basin management initiatives. These include the following:

● Operations studies on the three large reservoirs to develop multipurpose operating criteria and assigning top priority to flood control;
● More catchment rainfall gauging stations and improved flood forecasting;
● Flood control planning;
● Surveys and research on flash flooding;
● Flood warning and rescue systems.

---

[5]   This value was estimated based on the average irrigation application values from the farm household survey applied to the projected crop mix in 2010.

10.2.4.4 Domestic and industrial water supply

The Dong Nai basin is the economic centre of the country, and domestic and industrial water use has increased rapidly over the last decade driven by increased investments in industrial zones and water supply infrastructure. Despite this development, less than half of the population in the basin is served with public connections. For example, in 2000, the HCMC water supply company had a capacity of 837,000 m³/day (10% of which was from groundwater). Domestic deliveries in the same year reached about 417,000 m³/day for an urban population of 4.2 million people. Currently HCMC employs two large treatment plants utilizing water from the Dong Nai River. A third plant, on the Sai Gon River was completed in 2004. In addition, private wells accounted for estimated additional 400,000 m³/day. In general, urban areas such as Tay Ninh Town, Thu Dau Mot, Da Lat, Bien Hoa and HCMC have piped distribution systems for treated water drawn primarily from surface sources. Groundwater supplements these sources and provides the only source of supply for some smaller towns and many individual households.

Overall, domestic water demand has been estimated at 0.5 BCM, based on a standard of 160 l/day per person for urban areas and 60 l/day per person for rural areas. This probably exceeds current actual supply and use rates.

Industries in the basin are concentrated in the lower basin provinces, including Binh Duong, Dong Nai, and HCMC. Estimates for industrial use can vary by more than an order of magnitude, ranging from 162,000 m³/day based on delivery information to 2,000,000 m³/day based on an estimate of water use in the production of main industrial products in the basin. Industrial supplies come both from municipal-treated water supply systems and from individual river intakes or private wells. Organized industrial estates in some areas provide industrial water supplies from wells to concerns within the estates. Current industrial demand is estimated at just under three-quarters of a billion cubic metres annually and is expected to continue to grow rapidly.

10.2.4.5 Environment

Environmental uses of water in the basin appear, in practice, to have a lower priority than the extractive uses of irrigation and municipal and industrial water supply. Guidelines exist which suggest minimum flows of 25 m³/s at the future Ben Than intake on the Sai Gon River and 85–100 m³/s at the Binh An water supply intake on the Dong Nai River. These flows typically require releases of 25 m³/s from Dau Tieng, 100 m³/s from Tri An and 70 m³/s from the future Can Don reservoir. The Dau Tieng release satisfies needs of local irrigation on about 5000 ha in Binh Duong Province along the Sai Gon River below the reservoir and to control salinity, and the Tri An releases are to prevent the tidal salinity front from migrating upstream past the point of the HCMC municipal water intakes. At the same time, however, these releases have the effect of putting a small assured flow into the rivers below the reservoirs.

10.2.4.6 Other uses

Fisheries and aquaculture are relatively small activities in the freshwater portions of the basin, but are important contributors to local incomes and local diets. Cage aquaculture contributes importantly to waste loads in reservoirs, particularly Tri An.

The Sai Gon River is navigated by ocean-going freighters visiting the HCMC port. The rivers upstream from HCMC are used only by much smaller craft, and water flows and channel depths for navigation have not been a significant issue for basin management or reservoir releases.

10.3 Institutions and Actors

10.3.1 Legal framework

In 1999 and 2000, Vietnam initiated a series of major reforms in the country's water sector – including the framework Law on Water Resources (hereafter called the Water Law) of 1999, and the Decision on the Establishment of the National Water Resources Council (NWRC) in June of 2000. Among other things, the Water Law enables the establishment of river basin committees or organizations in the country.

These measures are meant to bring more coherence into a sector currently comprised of highly fragmented water authorities with sometimes overlapping responsibilities and little coordination. They follow the 1995 merger of the Ministry of Water Resources and the Ministry of Agriculture, which tended to align water resource management with one of its most important user sectors, but at the cost of potential for a broader resource-based perspective. As a result, the country is currently transitioning from a water sector that is fragmented and weakly coordinated to a more holistic, decentralized and integrated management of the country's water resources at the river basin level.[6]

The Water Law was adopted on 20 May 1998 and went into force on 1 January 1999. According to the law, water resources belong to the people under the management of the State, and organizations and individuals have a right to exploit and use the resources (Official Gazette, 1998). Water allocation is carried out from a river basin perspective adhering to the principles of fairness and reasonability. Priority in use is accorded to drinking water in both quality and quantity (Article 20).

According to the Water Law, the Ministry of Agriculture and Rural Development (MARD) is in charge of overall management of the country's water resources, but the government may delegate authority for specific water uses to other ministries. Water management is to be carried out in river basins that follow hydrologic catchment boundaries, and not administrative ones. MARD, together with provincial governments, is

in charge of establishing both flood and drought plans for the country's river basins. Both water use and wastewater discharge are to be licensed by provincial government authorities (People's Committees) under the guidance of MARD. Decree 179/1999/ND-CP of 30 December 1999 assigns specific duties to MARD, other ministries and Provincial People's Committees (PPCs) related to water resources management. Additional regulations are currently being drafted to implement the framework Water Law.

In addition to the Water Law, several other laws and regulations are important for water resources management in Vietnam. They include the Environmental Protection Law (27 December 1993) and the Ministerial Instruction from the Ministry of Science, Technology and Environment (MOSTE) entitled *Guiding Environmental Impact Assessment for Operating Units (Instruction No. 1420/QD-MTg)*.

In June 2000, a national-level umbrella organization for the water sector, the NWRC, was established, based on Article 63 of the Water Law (Government Decision No. 67/2000/QD-TTg). The NWRC has an office in MARD and a number of permanent members who represent the range of ministries and organizations that are involved in water resources management in the country.

In analysing water resource institutions, it is important to understand one salient concept pervading governance in Vietnam – the concept of *State management*. State management gives pre-eminence to the State across a wide range of public decisions. The key actor in this is the government, which is embodied in the office of the Prime Minister. Ultimate State management responsibility resides here, and the most important decisions are made by this office. State management authority for lesser decisions is delegated to ministries at the national level and to PPCs at the provincial level. For example, for water resources, Article 57 of the Water Law assigns a number of

---

[6]   In August 2002, after this chapter was written, water resources was again separated out from agriculture and placed under the newly created Ministry of Natural Resources and Environment (MONRE). Hence most of the functions ascribed below to MARD are now functions of the MONRE.

important State management functions to MARD. At the same time, Article 58 reiterates the supremacy of the government in exercising these State management functions and makes MARD answerable to the government in carrying out the responsibilities assigned to it.

A variety of other organizations, which do not possess State management authority, are placed under the direct supervision of an organ of government which does have it. Notable among these is the plethora of public 'companies', which have been created by various government units to provide particular services, such as infrastructure design and construction, or irrigation system management. These companies typically have strictly limited autonomy and are directly supervised by an entity with State management authority. This supervision goes beyond what is generally considered 'regulation' and blurs the distinction between government and nominally autonomous State corporations. While the concept of State management, as applied to State companies, appears to be evolving slowly to allow more autonomy, it is still a strong and integral part of their governance.

10.3.2 Basin actors and stakeholders

Actors can be defined as organizations and individuals who take part in the process of managing basin water resources. *Stakeholders* can be defined as those organizations and individuals having some direct interest affected by the outcomes of basin management decision making. The set of stakeholders thus includes actors, but is broader, in that it can include groups such as coastal fishermen, urban environmentalists and small farmers who may not have a direct role in basin management.

In Vietnam, actors in basin management tend to be government organizations rather than associations of individuals, trade organizations and other civil society groups. In the case of multi-province basins, government organizations at national, regional and provincial levels are typically involved,

with international agreements coming into play for the Mekong River. There are no international agreements for the Red River, despite its being an important trans-boundary river. In the case of the Dong Nai, the small portion of upper watershed located in Cambodia has so far not been cause for international consultations on basin management.

There is a strong sense of hierarchy in Vietnamese government structures, with bonds within ministerial groups of units and organizations being significantly stronger than those among units across different ministries. This is true even where these units are addressing a common set of issues. Within ministerial groups, there are often extensive consultations and upward requests for clearance before decisions are made. This can slow decision taking and limit cross-ministerial discussion. At the same time, it leads to coherence and conformity with ministerial policy directives.

The general form of public administration in Vietnam employs a matrix model, in which sub-national administrative units are responsible in two directions (Fig. 10.5). At the provincial and district level, sector-based units such as *Agriculture and Rural Development* or *Public Works and Transportation* are accountable administratively to the Provincial or District People's Committee, while in technical matters they are responsible to the corresponding ministry in Hanoi. Budgets flow through the People's Committee, and staff are employed by the province but receive technical direction from Hanoi.

This picture is complicated by the presence of a large number of 'companies', which have been created by government ministries and departments at both national and provincial levels. These companies are State-owned and typically closely tied to their parent ministry or department, often sharing staff, functions and budgetary resources. Most exist to do business with the government or with other State-owned companies.

In the water sector, important examples are the MARD engineering design companies, of which there are two at the national

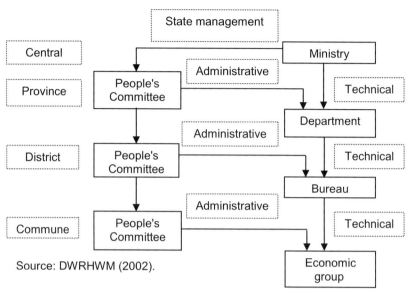

Source: DWRHWM (2002).

Fig. 10.5. Generalized management structure in Vietnam. Source: DWRHWM (2002).

level – Hydraulic Engineering Companies 1 and 2 (HEC 1, HEC 2) – and numerous ones at provincial levels, and water resources construction companies at both national and provincial levels. Similar sets of design and construction companies are found under the Ministry of Construction (MoC) and provincial construction departments for domestic water supply and under the MoI for hydropower development.

In addition, MARD and many provinces have set up irrigation management companies (IMCs) to supply water for irrigation and maintain irrigation facilities. Three of these are directly under MARD – the Dau Tieng Management Company, which operates Dau Tieng Reservoir on the Saigon River, and two others, which operate large irrigation systems in the north. Most provinces have also created one or more IMCs to operate irrigation systems within their boundaries. Provincial IMCs are units of the Provincial Agricultural and Rural Development Service (PARDS). The IMCs are generally intertwined with the PARDS themselves.

The government is moving slowly to privatize these engineering design and construction companies, and to put them into competition with a growing number of private firms in the same field. Currently, however, private firms are largely limited to subcontractor roles and many functions are still reserved for State firms.

Because IMCs operate in a natural monopoly setting, such competitive remedies are not available for the management problems affecting them. Moreover, some IMCs have also become involved in the construction business, creating further questions of conflicting interests and privileged competition against private firms. Future remedies lie in creating a much clearer separation between the parent government unit and the IMC, providing for user governance control of IMCs, transparent subsidies (if present) and other measures to enhance corporate behaviour and operating efficiency.

10.3.2.1 National and regional actors[7]

Important national and regional actors in the Dong Nai basin are shown in Box 10.1. The *government*, as discussed in an earlier

---

[7]   Descriptions of line government ministries and departments draw on DWRHWM (2002) and Ringler *et al.* (2001).

Box 10.1. National actors in water resource management of the Dong Nai Basin.

- Government of Vietnam
- NWRC
- MARD
  - DWRHWM
  - Various construction-related departments
  - SIWRP
  - Southern WR Research Inst
  - HEC 1, HEC 2
  - Construction companies
  - Dau Tieng Management Company
- MoI
  - EVN
    - NRC
    - PECC no. 2
  - Groundwater Department
- MOSTE
- GDHM
- MoC
  - Design Company for WSS
  - WSS Companies 1 and 2
- MoPH
- Ministry of Fisheries

section, exercises State management functions, in part through delegation to ministries and Provincial and District People's Committees.

In addition to MARD, the lead ministry, other ministries are also involved in the water sector, as delegated by the government. Currently, the water sector in Vietnam is in a transition period and water resources are still largely administered on a sectoral basis. Thus, for example, the MoI is responsible for the National Hydropower Plan; the MoC is responsible for urban water supply planning; MARD is largely focused on irrigation sector development and flood control; and MOSTE is responsible for water quality.

NWRC   The recently established NWRC is the apex organization for water resources in the country. The Council is chaired by a Vice Prime Minister and includes the Minister of MARD as a standing member and 15 other permanent members comprising vice ministers from Agriculture and Rural Development; Planning and Investment; Construction; Science, Technology and Environment; Finance; Transportation;

Fisheries; and Industry as well as the head of the General Department of Hydro-meteorology; the Chairman of the NWRC Office; and four specialists working in the water sector. The Chairman of NWRC may also appoint non-permanent members representing central and local agencies relating to specific issues on the agenda for each session of the Council.

The Council has the objective of facilitating coordination among the various ministries and agencies involved in water resources management. Its role is to advise the government on important decisions related to water resources management, including: (i) strategies and policies on national water resources; (ii) major river basin plans; (iii) plans for major inter-basin diversions; (iv) projects for protection, exploitation and utilization of water resources and projects for flood control and water damage control; (v) management, protection, exploitation and utilization of international water sources and dispute settlement; and (vi) conflict resolution between ministries and branches, and between ministries and provinces and cities under central control.

MARD   MARD, established in 1995 out of the three former ministries of water resources, agriculture and food industry, and forestry (Decree 73/CP of 1 November 1995), is the State agency in charge of water resources management and reports directly to the government. The 1999 Water Law reaffirmed this role.

MARD manages water resources in the country, especially in the fields of irrigation, drainage, flood control and bulk water supply. It is responsible for the overall policy framework, the planning and prioritization of new development, and the allocation of inter-provincial water resources. Funding for large capital projects, including main canals of large irrigation and flood control projects, is largely provided by the central government and includes significant foreign assistance. Secondary works and local projects are designed and funded by the provincial government with assistance from the central government.

Key water resources departments are: (i) Water Resources and Hydraulic Works Management (DWRHWM); (ii) Flood Control and Dyke Management; (iii) the Centre for Rural Water Supply and Sanitation (CERWASS); and (iv) Plan and Planning. Other associated units comprise design companies, construction companies, the Institute of Water Resources Planning, the Water Resources University and the Institute of Water Resources Research.

MARD is responsible for constructing dykes, headwork and canals down to 150 ha, and provinces are responsible for the areas below 150 ha. Operation and maintenance of existing irrigation systems generally rests with the provinces in which the systems are located. MARD controls three large inter-provincial irrigation systems – Bac Hung Hai, Bac Nam Ha and Dau Tieng.

MINISTRY OF PLANNING AND INVESTMENT (MPI)
The MPI is the central agency for allocating resources among sectors. It receives sectoral submissions and prepares consolidated public investment plans. The MPI coordinates Official Development Assistance (ODA) and direct foreign investment.

MOC   The MoC is responsible for urban water supply, drainage and sanitation. It sets regulations, and plans, designs and constructs water supply and sanitation facilities, working through associated design and construction companies. Following construction, management is transferred to a State-owned water supply company. The MoC is also responsible for water supply in small rural towns.

MINISTRY OF PUBLIC HEALTH (MOPH)   The MoPH is responsible for monitoring water quality and for sanitation.

EVN   EVN is responsible for hydropower development under the direction of its parent ministry, the MoI. It identifies hydropower development and establishes operating rules for reservoirs. EVN designs, constructs and operates schemes through associated companies.

MOSTE   The MOSTE was created in conjunction with the enactment of the environmental protection law in 1993. It sets water quality standards, conducts research and manages the environmental impact assessment (EIA) process for new water resource projects. Its primary action arm with respect to environmental management is the National Environmental Agency (NEA).

GENERAL DEPARTMENT OF HYDROMETEOROLOGY (GDHM)   The GDHM reports directly to the government. It collects primary level hydrometeorological data through its hydrology and hydrometeorology stations network across the country, conducts surveys, and provides analyses and forecasts based on the data it collects.

10.3.2.2 Provincial actors

The PPC is the paramount authority at the provincial level. Beneath it is a structure of departments that mirrors that existing at the national level. As indicated earlier, there are technical accountability linkages from these provincial departments back to corresponding ministries in Hanoi.

In the water resource sector, the PARDS is responsible for planning and implementing smaller water resource projects (less than 150 ha) and for irrigation system organization and management through associated IMCs. Other water-related responsibilities are carried out by departments generally corresponding to the ministries described above.

A number of national agencies have branches that are responsible for different regions of the country. However, branches of different departments and ministries have different geographic mandates. The result is that particular provinces may fall into a 'southern' region for one function and into a 'central' region for another. Table 10.3 shows the way that Dong Nai basin provinces are allocated among different sub-national jurisdictions for selected agencies. The result is an even greater number of actors involved in basin planning and management functions.

Table 10.3. Regional responsibility breakout by selected agencies.

	GW S	GW C	IWRP	SIWRP	FIPI S	FIPI N	NIAPP S	NIAPP C
Binh Duong	X			X	X		X	
Binh Phuoc	X			X	X		X	
Dak Lak		X	X			X		X
Dong Nai	X			X	X		X	
HCMC	X			X	X		X	
Lam Dong		X		X	X		X	
Long An	X			X	X		X	
Tay Ninh	X			X	X		X	

GW, Division of Hydrogeology and Engineering (working on groundwater); FIPI, Forestry Inventory and Planning Institute; NIAPP, National Institute of Agricultural Planning and Projections. S, south; N, north; C, centre.

10.3.2.3 Civil society

Civil society organizations, such as trade and professional groups, commodity producers, environmental and other issue-based non-governmental organizations (NGOs), and landowner groups are non-existent or relatively unimportant in Vietnam, at least in so far as water resources and agriculture are concerned. Basin planning and management organizations in closing (water-short) basins in other nations often include extensive participation of such civil society groups. These groups give voice to the interests of particular groups of stakeholders such as farmers, urban water supply utilities or the environment. Stakeholder interest groups are typically supported by member contributions and membership cuts across administrative boundaries. The relative absence of such stakeholder representation in Vietnam is a factor that must be taken into account in designing basin management organizations and considering their future development. Also to be considered is the political imbalance which may arise as powerful interests which are sophisticated and with relatively few members, such as industry groups, emerge in the absence of groups representing more numerous but less powerful and sophisticated stakeholders, such as agricultural water users. The natural environment represents a special case of particular importance in that it has no direct representatives and its interests must be advanced by public agencies such as MOSTE, or by nascent civil society groups such as environmental NGOs.

10.3.2.4 Dong Nai Basin Council

According to the 1999 Water Law, water resources in Vietnam are to be managed at the basin level. A Planning Management Council for the Dong Nai River Basin was established by a 9 April 2001 Decision issued by MARD. The Council consists of 12 standing and ten non-standing members, with the Vice Minister, MARD as chairman and the Director of the MARD DWRHWM as vice chairman. Standing membership also includes the Director of the SIWRP in HCMC, who is designated as Head of the Council Office. The Office functions as a secretariat for the Council and is housed at SIWRP in HCMC. Though the Office is not yet operating, two SIWRP staff members have been designated to staff it, and it has received a small allocation of VND 80M (US$5,333) to purchase furniture and office equipment. At the same time, councils were established for the Red River basin and the Mekong Delta. These three basins are to serve as pilot sites for implementing the basin planning and management concept stipulated in the Water Law.

Of the three possible models for basin organizations posited by Millington (1997), (i) a coordinating committee or council; (ii) a river basin commission; and (iii) a river basin authority, the council fits the first, and simplest, model, though it also contains a secretariat, which is characteristic of a basin commission in Millington's framework.

Significant features of the Council, as established, include the following:

- All 12 standing members of the Council are representatives of departments and ministries of the central government. Representation is generally at the departmental Vice-Director level. Ten of the 12 are based in Hanoi, while two are located in HCMC but directly connected with central government organizations.
- The ten non-standing members of the Council represent the ten provinces included in the planning basin (see *Basin boundaries* above).[8] Representation is at the relatively junior level of Vice-Director of the PARDS. As such, provincial representation does not include sectors of public health, domestic water supply, wastewater handling and treatment, and environmental quality. Non-standing members participate in the Council upon individual invitation.

Given their administrative level and status as non-standing members, provinces will likely have limited influence in the Council relative to representatives of the central government.

A second separate basin-level effort related to water quality management is also underway. This effort grows from a United Nations Development Programme (UNDP)-sponsored programme, which brought together representatives of Departments of Science, Technology and Environment (DOSTEs) in five key basin provinces in the late 1990s to discuss basin water planning and management. Although the original UNDP support for this activity has lapsed, provinces have continued to discuss a basin-wide effort focused on water quality management. A meeting in April 2002 resulted in a formal proposal for an interim

committee comprising ten basin provinces and HCMC, which was presented to the Office of the Prime Minister for decision. The interim committee would focus on:

- Developing a common information system for the basin;
- Increasing cooperation among provincial leaders for water quality management;
- Establishing communication channels among provinces for this purpose.

The proposal envisions primary representation by the provincial DOSTEs, a secretariat that rotates among the provinces, and the designation of an existing institution to manage a monitoring system for the basin.

10.3.3 Functions and actors

Chapter 1 described a set of essential functions of river basin management. A listing of these functions, crossed with the key actors identified in the previous section, is shown in Table 10.4. These functions are replicated, as appropriate, across three broad categories – surface water, groundwater and return flows (drainage). The return flow category includes both urban and industrial wastewater flows and agricultural return flows. Cells are marked to indicate an actor active in a particular functional area and whether the actor plays a major or a minor role in performing the function. Information is drawn primarily from informal focus group sessions with knowledgeable personnel at SIWRP combined with the experience of the authors. It is very important to note that the classification is based on assessments of actual effective performance rather than nominal or official responsibility. Thus, where a particular agency is given a clear official mandate to perform a function, but is ineffective in carrying out this mandate, its scoring

---

[8]   Long An, which – from a hydraulic perspective – has about 95% of its area within the Dong Nai basin, has not been included as a non-standing member, compared with Dak Lak province, which has only about 15% within the Dong Nai basin or Ninh Thuan and Binh Thuan with even lower shares.

Table 10.4. Key actors and functions matrix.

Key actors	Surface water									Groundwater							Drainage					
	Plan and prepare investments	Allocate water	Construct facilities	Distribute water	Maintain facilities	Monitor quality	Ensure quality	Protect against flooding	Protect ecology	Plan and prepare investments	Allocate water	Construct facilities	Withdraw/distribute	Maintain facilities	Monitor quality	Ensure quality	Plan and prepare investments	Construct facilities	Operate facilities	Maintain facilities	Monitor quality	Enforce quality
National																						
Government of Vietnam	●	●								○	○											
National Water Resources Council	○	○																				
MARD																						
Dept of WR and Hydraulic Works Management	●	●	○					○		●	●						○	○				
HEC 2	●	○															○	○				
Construction Companies			●		○							○						●		○		
Dau Tieng Mgt Company			○	○	●			●	○									○		●		
Dept of Dyke Mgt and Flood Control	○							●														
Dept of Investment and Construction	●		○															○				
Dept of Planning	●									○												
Dept of Policy	●									○												
Construction Mgt Board	○		○		○																	
International Cooperation Department																						
Sub-Inst for WR Planning	●	○				●		○		○	○						○					
Southern Institute for Water Resources Research	●	○				○		○		○	○						○					
Ministry of Industry										●												
Groundwater Department										●					●							

continued

Table 10.4. *Continued.*

Key actors	Surface water									Groundwater							Drainage					
	Plan and prepare investments	Allocate water	Construct facilities	Distribute water	Maintain facilities	Monitor quality	Ensure quality	Protect against flooding	Protect ecology	Plan and prepare investments	Allocate water	Construct facilities	Withdraw/distribute	Maintain facilities	Monitor quality	Ensure quality	Plan and prepare investments	Construct facilities	Operate facilities	Maintain facilities	Monitor quality	Enforce quality
EVN																						
Power Eng and Consulting Company no. 2	●	●	○																			
National Regulation Centre	●	●						●														
Construction Company			●		●																	
Equipment Installation Company			●		●																	
MOSTE																						
National Environmental Agency	○					○	○		○	○					○							
Gen Dept of Hydromet no. 2	●					○	●	○		○					○		○					
Ministry of Construction																						
WS Design Company no. 2	●	●	○							○	○						●					
WS Construction Company			●		●							●		●				●		●		
Ministry of Public Health						○	○															
Ministry of Fisheries	○																					
Ministry of Transportation	○																					
Ministry of Defence (Environ Protection Centre)						○															○	

Provincial

- PPCs
- PARDSs
- Sub-Dept of WR and HW Mgt
- Irrigation Mgt Companies[a]
- WR Consulting Companies[a]
- WR Construction Companies[a]
- Dept of Dykes and Flood Protection
- DOSTEs
- Department of Transportation
- Dept of Construction
- Water Supply Companies[b]
- Drainage Companies[b]

Regional & Special

- HCMC Export Processing and Industrial Zones Authority
- Dong Nai Basin Committee[c]
- Private industries
- Pvt consulting and construction firms
- Universities and other institutes
- International consulting firms

[a] Directly under the PPC in some provinces with broader mandate.
[b] Combined in some provinces.
[c] Not yet functional; anticipated initial roles.
●, Major role; ○, minor role.
MARD, Ministry of Agriculture and Rural Development; HEC, Hydraulic Engineering Company; EVN, Electricity of Vietnam; MOSTE, Ministry of Science, Technology and Environment; PPC, Provincial People's Committee; PARDS, Provincial Agricultural and Rural Development Service; DOSTE, Department of Science, Technology and Environment; HCMC, Ho Chi Minh City.

may indicate that it plays a minor role, or no role at all. Different observers can obviously score the matrix in different ways, depending on their assessments of performance and effectiveness and the weights they attach to different components of performance.[9] Nevertheless, the matrix gives a broad indication of which functions are being covered effectively, and by whom, and which are not being addressed or are addressed by a multitude of actors. It could readily be refined – by developing separate tables for nominal and actual responsibility, for example – but is introduced here to serve as a base for discussion and a broad indication of involvement and responsibility. Descriptions of the functions shown in the text are given in Chapter 1.

All functions represent the 'action edge' of functional performance. Underlying each are a set of supporting functions such as data collection and management, resource mobilization, consultation and so on, that are required to carry out the top-level function shown. For simplicity, these have not been shown separately.

10.3.3.1 Key functions

Three key functions are of greatest importance for basin level water management in the Dong Nai basin. These are resource development, water allocation and water quality management, and are discussed below. Other functions, particularly those relating to water service provision, are also important, but are beyond the scope of this chapter (World Bank, 1996; Ringler *et al.*, 2001; Chapter 1 of this book).

RESOURCE DEVELOPMENT   Resource development remains the primary focus of water resource management in Vietnam. Important functions involved are *planning and preparing investments*, and *constructing facilities*. A very large number of ministries,

departments and institutes are involved in the planning and investment preparation process, particularly for surface water resources. Many of these entities lie within MARD itself. Elaborate procedures for this survive from the earlier era of comprehensive central planning. Many of these are based on old Soviet standards and procedures, but are being slowly updated to achieve compatibility with standards and procedures used by multilateral lending institutions and other international lenders.

Construction of water resource facilities is carried out principally by State-owned construction companies, and a number of State agencies are involved in monitoring and supervising construction. For smaller projects, planning and construction processes employed at the national level are replicated, in somewhat simpler form, at the provincial level.

Important resource development decisions are ultimately made by the government or PPCs.[10] In addition to their significance for water resource development, these decisions also have important implications for water allocation, as discussed in the following section.

WATER ALLOCATION   Water allocation can be accomplished in different ways. These range from a permitting or rights allocation system covering all uses, to an open access system where water is available for the taking. The 1999 Water Law calls for a permitting system for both surface and groundwater to be operated by MARD, granting use rights for water to larger-scale users. Rights to groundwater use would be granted for 15 years and surface water use rights for 20 years. Small quantities of both ground and surface water may be extracted and used without permission by individual families for domestic use and for small-scale agriculture, forestry, aquaculture and other home

---

[9]   This makes the matrix a useful tool in group planning processes to illuminate strengths and weaknesses, as well as areas of agreement and disagreement among stakeholder participants.
[10]   The MPI plays an important role in allocating investment funds among projects, but is not separated out here for simplicity.

enterprises. During periods of shortage, priority is accorded to domestic water supplies over all other uses. MARD is currently developing regulations for implementing this important provision of the law, but the permitting system is not expected to be operational until 2003.

Allocation of access to surface water use is currently controlled implicitly through the investment decision-making process. Since major users of water are almost exclusively public agencies and companies, public investments in new water-using facilities determine, *de facto*, the basic allocation of surface water among users. Such uses include irrigation schemes, hydropower stations, riverine port facilities and municipal water supply systems. Obviously, the State would not construct these facilities unless it intended that they have access to a reliable supply of water – a *de facto* allocation of water use rights. Hydrologic studies conducted during project preparation determine the availability of water in the source to be tapped, and may also detail the expected return flow contribution to receiving waters. By preparing plans for such facilities, planning agencies such as SIWRP, HEC 2 and PECC 2 play important supporting roles in water allocation. Ultimately, water-allocating investment decisions are taken by the government, on the advice of MARD and its DWRHWM.

Surface water can easily be over-allocated in this way as basins approach closure; hence the need for a more rigorous and explicit allocation system as envisioned in the Water Law. Conflicts among water uses can also arise over the shorter term, during periods of seasonal low flow and during drought years. Similarly, the need to control flooding creates the need to make allocation decisions – between hydropower generation and increased flood storage capacity, for example – over the shorter term.

In addition to the allocation of longer-term access to the water resource, short-term allocation decisions are required to deal with changes in water availability and demand throughout the year and to allocate

shortages in low flow years. In these cases, where capacity has been created to use more water than is available at a given time, some process is needed to allocate the scarcity among existing users. The Water Law indicates that in such cases, after giving priority to domestic water supplies, water shall be distributed according to 'the percentage defined in the planning of the river basin and the principle of ensuring fairness and reasonability'. EVN is an extremely important actor in short-term allocation through its decisions about when and where to generate power. The timing and magnitude of power releases have the effect of controlling the amount of water available to downstream users, particularly during the dry season.

Presently decisions regarding the allocation of shortages among users during droughts and seasonal low-flow periods rest with the Southern Disaster and Flood Committee. However, this committee is mostly concerned with flood prevention. In practice, shortages are allocated by PPCs for reservoirs located within a single provincial jurisdiction. At present, there does not appear to be a comprehensive functional system for dealing with shortage conditions.

With respect to flood control allocation decisions, dam managers, i.e. the Dau Tieng Management Company and EVN, make release decisions on the basis of unchanging rule curves as discussed elsewhere in this chapter. These allocation rules are thus largely fixed at the time of dam design, though operators have some operational flexibility over the short run within the confines of the upper and lower boundary curves. In the case of severe floods, the Southern Disaster and Flood Committee also meets and has the power to change allocations.

A more formal allocation procedure is currently in place on an interim basis for groundwater. These interim procedures will be supplanted when implementing regulations for permitting under the 1999 Water Law are finished. Present rules call for groundwater abstractions greater than 1000 m³/day to be licensed by MARD, while smaller abstractions are licensed by provincial authorities. Permits are not required for minor abstractions.[11]

WATER QUALITY MANAGEMENT   Water quality management consists of two basic functions – quality monitoring and quality assurance and enforcement. The former consists of data collection, processing, analysis, dissemination and storage. The latter is a more complex task consisting of standard setting, problem recognition, problem diagnosis and problem solution through provision of incentives, enforcement of standards and regulations, and mitigation of damage.

Both monitoring and quality assurance functions can be applied to ambient waters, both ground and surface, and to drainage discharges from potentially polluting water uses, including agriculture, municipalities and industry. In all cases, it is important to address both the simple monitoring function and the more complex and difficult quality assurance and enforcement function.

In the Dong Nai, a number of organizations from both national and provincial levels are active in monitoring the quality of surface water in basin rivers. These include the NEA, MoPH, SIWRP, Southern Institute for Water Resources Research (SIWRR), GDHM, DOSTEs and water supply companies. Groundwater quality is also monitored by several agencies and departments, though less extensively than surface water. Monitoring of drainage water quality, particularly municipal and industrial point source return flows, is more limited still. Overall, there are problems with inconsistent data sets, lack of continuous time series data, and a profound reluctance to share data and information with other agencies and with the general public. Monitoring is taking place, but it is fragmented and sometimes duplicative, and information is difficult to access. Most basin provinces feel the need for an integrated water-quality monitoring network for the basin.

There is far less activity in the area of ensuring quality of receiving waters and enforcing water quality standards. The provincial DOSTEs are the primary enforcing actors, but they often lack resources and clout to provide aggressive enforcement. At its present stage of economic and water resource utilization in the country, emphasis is on resource development, economic growth, expanding industrial output and employment, and enhancing agricultural exports. These objectives tend to take precedence over those related to water quality assurance. Concomitantly, the influence of national ministries and municipalities and provinces which host major industrial growth zones is generally greater than that of offices charged with protecting the environment. This results in a situation in which enforcement of existing quality and discharge standards is incomplete.

Fortunately, the general abundance of water in the Dong Nai basin has prevented pollutants in wastewater discharges from causing major problems thus far. Despite the fact that HCMC, a city of 5 million people, currently has no wastewater treatment facilities whatsoever, pollution in the coastal estuary of the Saigon River, which receives this effluent load, is said not to be a significant problem. Still, this is an issue likely to grow to significant proportions in the near future and will require cross-sectoral and inter-provincial cooperation to address.

10.4 Perspectives for the Future

Vietnam in general and the Dong Nai basin in particular remain in an evolutionary stage of water resource development. There is still relatively little pressure on the overall water resources in the Dong Nai basin, and water quality problems have yet to reach crisis proportions. Concomitantly, primary emphasis remains on increasing storage and developing new use potential in the basin rather than on improving efficiency in individual use sectors or on re-allocating water from lower to higher productivity uses.

---

[11]   In Binh Duong province, for example, this is interpreted to apply to wells smaller than 60 mm.

There are both historical and resource-based reasons for this. Vietnam did not participate in the reservoir construction boom, which took place in Asia during the 1970s and 1980s. As a result, there is presently storage for only about 13% of the annual discharge of the Dong Nai River system, and major portions of its large monsoonal flow reach the South China Sea unused. This limits dry-season water availability for all uses, as well as the options available for controlling and allocating water in the basin. Hence, while on an annual basis the basin is still amply supplied with water, seasonally a constraint is rapidly being approached.

A water quality problem is emerging together with growing dry-season scarcity. Burgeoning industrial growth and an absence of municipal wastewater treatment are degrading dry-season river water quality with consequences for users of high quality water such as HCMC and other expanding lowland population centres, and certain kinds of industries such as food processing.

Authorities are moving to address problems on several fronts. These include regularizing the permitting system for water use rights, monitoring water quality and establishing a basin planning committee. Overall, however, primary emphasis appears to rest on ambitious storage development plans for the basin, which call for a 35% increase in surface water storage over the next 8 years. However, the fact that much of this storage is being developed for hydropower production leads to important potential conflicts in operational goals among hydropower operators, irrigators, municipalities and the environment.

Presently, surface water allocation is implicit in the basin investment planning process. Individual projects are evaluated with respect to water supply availability and, once approved and constructed, possess a de facto right to use water. MARD is currently working on implementing regulations for a formal system of use rights allocation as called for in the 1999 Water Law. Groundwater regulations are being tackled first, to be followed by surface water permitting rules. Principles articulated in the Water Law establish priority for domestic water use and provide for a proportional system of sharing shortages, though details of this procedure remain to be worked out.

A basin planning management council for the Dong Nai was established in 2001, but operating regulations for the council are still being worked out. Those regulations will provide more insight into the functions the council will be performing, which, at present, are very broadly defined.

Based on the make-up of the council, the resource planning mandate of its secretariat, and the pattern of public priorities and decisions observed to date, it is possible to speculate that the primary role of the council, at least at the outset, will be in planning new basin development activities. In this role, the council will serve as an important consultative body for SIWRP in the planning process.

Other roles may be added subsequently as basin infrastructure becomes more fully developed, stress on the available resource increases, and water quality deteriorates. A primary requirement for more comprehensive basin management will be the devolution of greater power over resource management to provincial authorities. This is happening, but at a very deliberate pace. Provincial, municipal and other local authorities are closer to the needs and problems of the water users in the basin and are in a better position to represent those interests. They also have more at stake in satisfying local constituencies. Ultimately, one would expect to see direct representation emerging, as interest groups in civil society develop their own structure and voices. Over the short-term, it is likely that important decisions with respect to basin management will continue to rest with MARD and the MoC and EVN in Hanoi, and that decisions will be more grounded in national interests than in regional or local ones.

A parallel basin-wide coordinating effort has been ongoing for several years under the sponsorship of MOSTE and support by the UNDP. This effort differs from the MARD-based planning management council in that it is based in provincial governments and in its focus on water quality issues. Ultimately, these two efforts will need to be integrated or closely coordinated. For the time being, they appear positioned to evolve on parallel tracks. In some ways, this is a positive development as it aids the emergence of a provincially based capacity to engage in basin management activities. Coordination will be required and, at some point, the two mechanisms may be merged, but this should not be done at the expense of the seemingly more democratic and representative process emerging at the regional level.

What are the implications for irrigated agriculture of this unfolding scenario? A principal one is that a management strategy based on expanding basin storage is unlikely to stress agriculture significantly, or require it to relinquish water supplies it is currently using, or even to curtail growth in the agricultural sector. Additional storage should enhance dry-season water supplies, which can be used by urban and industrial users downstream of the turbines of the hydropower dams. In addition, the position of MARD as both water resource manager and agricultural supervisor should assure a privileged place for allocations to agriculture in the basin.

Within the irrigated agricultural sector, strongest growth is expected in the cash-crop sector, where shallow groundwater is the most important source of supply. This has the important advantage of being widely available and being a demand-driven supply under the control of the well-owning farmer. On the other hand, shallow groundwater sources are vulnerable to drought-induced shortages, since the shallow aquifer reservoir volume is often small, and to over-exploitation. This leads to the need to consider strategies to replace reliance on groundwater with farmer-controlled reservoir-backed surface water supplies, and for artificial groundwater recharge to buffer the effect of seasonal droughts on high-input cash crops, particularly perennial crops.

References

DWRHWM (2002) *Water Resources Development and Management in the Dong Nai River Basin*. Report prepared for the ADB-IFPRI project Irrigation Investment, Fiscal Policy, and Water Resource Allocation in Indonesia and Viet Nam. MARD, Hanoi.

MARD (2000) Decision of Minister on Dau Tieng Reservoir Temporary Operational Regulation. No. 127/2000/QD-BNN-QLN, 18 December.

Millington, P. (1997) Institutional development options for the water management of the Dong Nai Basin. Unpublished report.

Ngoc Anh, N. (2000) Preliminary report. Flood situation in Dong Nai River system in 2000 and flood control plan (Bao Cao So Bo – Tinh Hinh Lu Lut Nam 2000 O He Thong Song Dong Nai Va Nhung De Xuat Ve Chien Luoc Kiem Soat Lu). Sub-Institute for Water Resources Planning (Mimeo).

Official Gazette (1998) *The Law on Water Resource* (No. 8/1998/QH10 of May 20, 1998). English Translation. Official Gazette No. 21 (31 July, 1998), pp. 32–47.

Ringler, C., Chi Cong, N. and Vu Huy, N. (2001) Water allocation and use in the Dong Nai River Basin in the context of water institution strengthening. Paper prepared for the workshop on *Integrated Water Resource Management in a River Basin Context*, Malang, Indonesia, 15–19 January. IFPRI, Washington, DC.

SIWRP (2001) Data provided to the ADB/IFPRI RETA 5866 project. SIWRP, Ho Chi Minh City.

World Bank (1996) *Vietnam Water Resources Sector Review*. World Bank, Washington, DC.

11 Governing Closing Basins: the Case of the Gediz River in Turkey

Mark Svendsen, D. Hammond Murray-Rust, Nilgün Harmancıoğlu and Necdet Alpaslan

11.1 Introduction

11.1.1 National context

Turkey is a rapidly modernizing country of about 66 million people with a population growth rate of 1.5% per year. At the beginning of the 1980s, national economic strategy switched from a policy of industrialization based on import substitution to a policy aimed at allowing a greater role for markets. Between 1979 and 1993, the Turkish economy expanded at an average rate of around 5% per year, with growth in recent years being even faster. The GNP in 1998 was US$200 billion, similar to that of Austria or Sweden. However, per capita income was a more modest US$3160, slightly less than those of Malaysia or Venezuela. Turkey is a very dynamic but still emerging country, displaying characteristics of both developing and developed nations.

The service sector dominates the economy, contributing 62% of GDP in 1999, while agriculture provided just 15%. However, agriculture still employs a very significant 45% of the nation's workforce. Agricultural incomes, though, average less than one-quarter of those in other sectors. Regional disparities in income and other measures of development also remain significant.

As a result, extensive internal migration is leading to rapid urbanization. At present, 65% of the population lives in urban areas, and the urban population is growing at 2.8% per year while the rural population shrinks at an annual rate of 0.7%.

Turkey clearly sees itself joining the ranks of economically developed Western democracies and hopes to join the European Union by 2023, the 100th anniversary of its independence. This is a powerful motive across all sectors for harmonizing Turkish policies, practices and standards with those of the EU.

At a national level, Turkey is a parliamentary democracy. At the sub-national level, the country is governed by a mixed system employing both local elections and central government appointments. Population centres (cities, towns and villages) are governed by locally elected assemblies or councils with administrations headed by locally elected mayors. Provinces and districts, while having locally elected assemblies, are headed by senior civil servants appointed by the Ministry of the Interior. National-level policy guidance and instruction is important at all levels and centralized revenue collection makes revenue transfers from national to local levels important and enhances the power of the centre.

11.1.2 The Gediz basin

The Gediz basin is located in Western Turkey (Fig. 11.1) near Turkey's third largest city, Izmir, which is, however, located just outside the basin.

The Gediz River rises in the mountains of Western Turkey and enters the Aegean Sea to the north of Izmir. It covers a distance of about 400 km and drains an area of 17,000 km², about 8% of which is irrigated. The basin had a population of 2.33 million in 1997.

The Gediz basin has changed considerably in the past decade, shifting from a comparatively water-rich basin to one that is now closing.[1] Change has been driven by an above average increase in urban and industrial demand, rapidly growing concern for issues of water quality and environmental protection, and reduced rainfall over the catchment. Paralleling these hydrologic changes has been a much slower institutional response that has not kept up with the requirements for changes in the way water is allocated and managed.

11.1.3 The case study

Characteristics of the nation and the basin together make the case an interesting one for study for the following reasons:

- Turkey is a dynamic middle-income country with rapidly growing demands for industrial and municipal water supplies;
- Agriculture is an employment source for nearly half of the Turkish population and the country's most important current user of water;
- The Gediz basin represents these trends well, containing both significant agricultural water use and rapidly expanding municipal and industrial demands; moreover, its water resources are nearly fully allocated.

Although the drought of the early 1990s has now passed, the legacy is seen in a number of important issues that continue to lie at the core of the debates surrounding the management of water in the Gediz basin. Several of these are highlighted below and discussed in more detail in the remainder of the chapter:

- The increasingly apparent need for a unified coordinating mechanism for water allocation among irrigation, urban areas, industries and the environment to replace existing bilateral processes;
- The continuing struggle between older long-established institutions dealing with water resource development and water allocation and emerging institutions concerned primarily with water quality and environmental issues;
- The need to represent and protect the interests of certain water users, such as the Gediz delta ecology and

Fig. 11.1. Location of Gediz basin.

---

[1]   A closed basin is one in which there is no unused water left to be allocated.

the Irrigation Associations (IAs) established during the past 5 years to assume O&M responsibility for 110,000 ha of large-scale irrigation systems, within the wider debate of water resources allocation and management;

- The need for: (i) clear rules assigning responsibility for setting water quality and quantity standards and monitoring actual conditions; and (ii) sufficient political power and will to sanction violators of the standards.

In the remainder of the chapter, we deal first with the hydrology and water use patterns within the basin and then turn to the legal, policy and institutional conditions which influence how the basin is governed and managed. Finally, we combine the two assessments to summarize the problems facing the basin and the challenges it must meet to overcome them.

11.2 Water Resources of the Gediz Basin

11.2.1 Hydrology

11.2.1.1 Surface water

The hydrology of the Gediz basin is typically Mediterranean. Precipitation falls between November and April, and peak river flows occur in February or March. Annual precipitation varies from 800 mm in higher inland areas to about 450 mm near the coast, with about 80% falling in the winter months. Under natural conditions, there is a steady decline in stream discharge until May when many of the smaller streams dry up. Summer flows are only present in the Gediz River and its largest tributaries, and even they may be negligible in the peak summer months. Following the irrigation season, the only flows in the Gediz River are from the few larger tributaries plus residual return flows from irrigated areas and industrial and municipal wastewater discharges to the river (Fig. 11.2).

Following the construction of Demirkopru Dam in the 1960s, net annual surface water availability in the main basin and the delta is estimated to have been approximately 1900 million cubic metres (MCM) per year. During the period 1989–1994, a severe drought affected the basin, and since 1990 there has been a persistent decline in surface water flows into Demirkopru. Annual water availability since that time has averaged just 940 MCM. As some of this flow occurs in winter and is derived from tributaries where there is no storage, there is little difference between annual surface water availability and current demand of about 660 MCM.

11.2.1.2 Groundwater

Groundwater resources are able to make up some of the potential shortfall in overall water availability. The central part of the basin is an alluvial plain whose groundwater reserve is replenished in most years.

In the alluvial fan areas on either side of the main valley and in the Nif Valley, the situation is more critical. Tubewell-based farmers in the Akhisar area complain of a steady and long-term decline in water tables, and in the Nif Valley, the water table is reported to have dropped by 5–8 m in the past 10 years as industrial extractions have burgeoned. Springs in the limestone areas are also reported to have declined in the past decade.

The estimated safe annual yield for groundwater in the main part of the valley is 160 MCM/year, which is about one-third less than the 219 MCM estimated as being extracted from the main and Nif valleys. Despite the absence of definitive figures, it appears that groundwater use presently exceeds, by a sizeable margin, the sustainable limit.

In the Gediz delta there is little groundwater utilization, and extraction near the coast is prohibited to prevent saltwater intrusion. The groundwater is deep and therefore there are no shallow tubewells.

11.2.2 Basin water use

The Gediz basin contains a typical range of water users, although the balance among them has been changing during the past

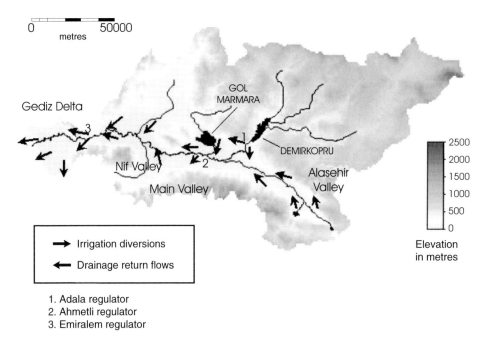

0 ————— 50000
        metres

Gediz Delta

GOL MARMARA

DEMIRKOPRU

Nif Valley

Main Valley

Alasehir Valley

Elevation in metres

2500
2000
1500
1000
500
0

➡ Irrigation diversions

⬅ Drainage return flows

1. Adala regulator
2. Ahmetli regulator
3. Emiralem regulator

Fig. 11.2.   Irrigation and drainage flow patterns, Gediz basin.

couple of decades. Each user category is described briefly.

IRRIGATED AGRICULTURE   Traditionally, the largest user of water has been irrigated agriculture, originally deriving from small run-of-the-river diversions from the Gediz and its tributaries dating back some 3000 years. Since 1945, the development of large-scale systems and groundwater exploitation has transformed irrigated agriculture.

LARGE-SCALE SURFACE IRRIGATION   The first investments in modern irrigated agriculture began in 1945 with the construction of two large regulators to tap the flow of the Gediz River. Adala regulator serves some 20,000 ha of land in the middle portion of the basin, whereas Emiralem regulator commands 22,000 ha in the Gediz delta (Fig. 11.2). In the 1960s, a second set of investments were made that included the construction of Demirkopru Reservoir a few kilometres upstream of Adala, a third regulator at Ahmetli, and the regulation and raising of the natural lake of Gol Marmara. Ahmetli Regulator commands some 45,000 ha of land. The final surface

water developments took place in the Alasehir Valley with the construction of two small reservoirs. The total command area of the large-scale surface systems is approximately 110,000 ha. The predominant crops are cotton (50%), grapes (35%), maize, fruit orchards and vegetables.

At present, surface water issues from Demirkopru are limited to the interval between mid-June and mid-September, which is focused on the cotton-growing season. Natural stream flows from tributaries can be used for land preparation for cotton or for early irrigation of grapes and fruit trees, but there are no releases made into the Gediz River from Demirkopru outside this period.

Water use in the 90,000 ha of the central and delta zones is limited to 75 m³/s from Demirkopru and 15 m³/s from Gol Marmara for a release period of approximately 60 days, or a total of some 550 MCM during the year. This is equivalent to some 450 mm of irrigation water for the growing season.

In the Alasehir Valley in the east of the basin, irrigation is almost exclusively for grapes. Application rates are approximately 350 mm/season and during the summer there is no significant net outflow into the main part of the basin. It is estimated that, through a combination of surface application and some pumping of the shallow aquifer, approximately 60 MCM are consumed during the summer season.

SMALL-SCALE SURFACE IRRIGATION   In many tributary valleys into the Gediz, there are small-scale surface water diversions that take advantage of winter runoff and spring snow melt. Typical crops are fruit trees, winter wheat and vegetables, because these only require water in spring and early summer before the streams dry up.

There are no accurate records of the total area involved, but it almost certainly is more than 25,000 ha, since almost every village situated on the valley fringe has some irrigated area (Kayam and Svendsen, 1999). Because the number of irrigations is low, normally two to four irrigations of about 50 mm each for the entire season, total water use is also low and is estimated at 50 MCM.

GROUNDWATER IRRIGATION   There are two different categories of groundwater users: those who are members of village or pump cooperatives and those who make private investments.

Starting in the 1960s, but increasingly in the 1970s, the government has fostered community-based irrigation from deep tube-wells. Most deep tubewells have discharges in the range of 50–150 l/s and are often tapping groundwater at least 100 m below the surface. The majority of wells are outside the boundaries of the surface irrigation systems and are concentrated in the Akhisar and Nif Valleys (Fig. 11.2). Typically, crops are high-value, and include tobacco, vegetables and fruit trees. Total water extractions are estimated at 30 MCM/year on the basis of 100 wells having a typical discharge of 75 l/s and operating for 40–50 days/year.

Private groundwater exploitation started during the drought of 1989–1994. Many individuals purchased centrifugal pumps to exploit shallow groundwater within the boundaries of surface irrigation systems, and in some cases neighbours formed informal pump groups to purchase a pump and well and then share operating costs at the end of the season. Farmers have continued using these pumps and there has been a recent small increase as some farmers adopt trickle irrigation systems for high-value fruits and vegetables.

The vast majority of private pump owners are within the boundaries of the surface irrigation systems. As such, they rely on the seepage and management losses from the surface irrigation system. While there is some evidence that the shallow ground-water table dropped during the drought, it has since recovered and it is assumed that they do not mine groundwater but merely re-use surface water. Their net water use is therefore included in the total surface irrigation volumes. However, official records of surface irrigation shows that only about 70% of farmers use surface water and some of those also pump. It is estimated that some 40,000 ha of land in the command areas of the surface irrigation systems are actually pump-based with only 70,000 ha relying primarily on canal water.

A few private pump owners are situated on the fringes of the surface irrigation system or in the area between the Alasehir Valley and the main Gediz Valley. These are estimated to use some 5 MCM/year that is not direct recharge from surface irrigation systems.

11.2.2.1 Municipal water supply

The Gediz basin has two separate classes of urban and municipal water users: the towns and villages within the basin itself and a substantial transbasin diversion of drinking water to Izmir.

WITHIN-BASIN USE   There are no accurate records of total water extractions for urban and municipal water consumption in the Gediz basin. All municipal extractions are from groundwater. Based on estimates provided by the different municipalities, it appears that extractions are in the order of 130 MCM/year. However, much of this returns in the form of wastewater as either percolation into the groundwater or discharge in surface water. Allowing for 20% actual consumption, the net municipal extraction within the basin is estimated to be 26 MCM/year.

Some municipalities also have shown interest in using good-quality spring water for their water supply. In a few cases, municipalities have arranged with villages to use a portion of their spring water and agree to compensate them by improving village irrigation systems.

IZMIR USE OF GEDIZ WATER   Izmir has had a long-standing claim on groundwater within the Gediz basin and there are two main well fields, at Sarikiz in the north of the basin and Goksu near Manisa. Actual consumption data are not available, but Izmir has extracted as much as 108 MCM/year from these well fields. Because the water is transferred out of the basin, there is no return flow.[2] An important potential source of additional water for in-basin use is the estimated 50% of the water entering the Izmir municipal system that is lost to underground leakage. Since Izmir is on the sea, no reuse of these losses is possible. If the conveyance efficiency of the Izmir piped system were improved, up to 50 MCM of high-quality groundwater could be left in the Gediz basin annually and used for other purposes.

11.2.2.2 Industry

There are two important industrial areas in the basin. The largest is in the Nif Valley immediately east of Izmir in Kemalpasa municipality (Fig. 11.2). There is also a growing industrial estate on the western edge of Manisa. Industries included ceramics, leather, food processing, metal works and assembly plants.

In both areas, groundwater is used for the industries, and each industry must obtain a permit from the General Directorate of State Hydraulic Works (DSI). However, there are no records of how much water

is consumed and it is difficult to make an estimate of total water use.

11.2.2.3 Hydropower

Between 1970 and 1988, Demirkopru Reservoir was used for hydropower generation throughout the year. Since the drought, however, power generation has been restricted to periods when water is released for irrigation and no special releases for hydropower are made.

11.2.2.4 Environmental consumption

The seaward fringe of the Gediz delta is an important nature reserve and has recently been designated as a Ramsar site to protect rare bird species. Originally, the area received excess water from the Gediz River for much of the year, but since 1990, with restrictions on irrigation releases, the reserve suffers from water shortages. The summer months are the critical time for providing water specifically for the nature reserve, since during the winter water is available from the Gediz River before flowing to the sea.

In response to demands to preserve bird habitat, one small channel with a capacity of 0.7 m³/s does now extend from the irrigation system into the nature reserve. However, the channel does not always flow at the maximum rate during the 60-day irrigation season, and so the potential volume of just under 4 MCM for the season is not normally provided. One preliminary estimate suggests that to maintain appropriate conditions for freshwater bird habitat, as much as 1.5 m³/s is required during the 120 days of the summer season, a total of about 15 MCM. Two other options are available for supplying water to the reserve. A pair of pumps on a nearby drainage channel can supply 0.5 m³/s and a well is available which can provide a modest 0.05 m³/s. These pumps

---

[2]   Plans are in place for Izmir to supply irrigators in lower portions of the Gediz basin with treated wastewater. Irrigators are enthusiastic about this because of the very poor quality of the surface water they currently receive. There are concerns, however, about possible high salinity levels in the treated effluent. At design output, the treatment plant would produce about 880 MCM/year, roughly equivalent to the entire current use in the basin.

have not been operated in recent years because reserve managers have not requested additional water.

A second component of environmental demand is the water needed for waste conveyance from points of origin within the basin to the sea. In transporting wastes, the flow must provide sufficient velocity to keep organic compounds and heavy metals adsorbed on to soil particles from settling out before reaching the sea and sufficient dilution to avoid in-stream environmental harm. Obviously, reducing the pollutant loads that must be carried would reduce the quantity of water needed for this purpose.

11.2.3 Summary

The total estimated water extraction by the different users is shown in Table 11.1.

Irrigation currently uses a large share of the surface water resources of the basin. Withdrawals total about 660 MCM, with 83% of that going to large-scale irrigation systems. Current surface water allocation practices are primarily aimed at providing reliable water deliveries to the IAs in the large irrigation systems, and this has been achieved with considerable success. Hydropower generation has no priority of its own and uses only water that is to be released for irrigation. A small and probably inadequate

allocation of poor-quality surface water is currently made for the wetlands in the Gediz delta.

Heavily polluted wastewater discharges from urban areas and industries within the basin seriously degrade the quality of surface water in natural channels, particularly in the low-flow summer months. Since water use for these purposes is growing at an estimated rate of 6–8% per year, this degradation can be expected to worsen unless major efforts at control are made successfully.

Groundwater supplies roughly a quarter of basin water use, of which about 16% is for irrigation and the remainder for urban and industrial use. Groundwater supplies nearly all of the water used for these latter two purposes. Irrigation use of groundwater is largely static or declining as less-water intensive crops replace cotton and improved water application technology gains a foothold. Municipal and, particularly, industrial use is expanding rapidly, however. At present as much as one-quarter of groundwater withdrawal in the basin may be unsustainable overdrafting, and pressure on these aquifers is expected to increase as industrial demand continues to grow.

Much of the water withdrawn for municipal and industrial use within the basin is returned to surface waterways, but in seriously degraded condition. This, in turn, gives rise to a need for additional

Table 11.1. Estimated water use by sector.

| Water user | Estimated consumption | | Notes |
	MCM	Share of total	
Surface water			
Large-scale irrigation	550	62%	From Demirkopru and Gol Marmara
Small-scale irrigation	60	7%	Alasehir Valley
Hydropower	50	6%	No priority for hydropower
Bird reserve	4	–	Current releases only; needs more
Groundwater			
Pump irrigation groups	30	3%	Only those outside surface irrigation area
Private irrigators	5	1%	18% of extraction, remainder is return flow
Urban within the basin	26	2%	Trans-basin transfer, no return flow
Transfer to Izmir City	108	12%	Estimated by DSI
Industry	50	6%	

MCM, million cubic metres.

allocations of surface water for waste load transport and dilution, water which is simply not available at present. The alternative is to improve quality of wastewater discharges significantly at their sources.

11.2.3.1 Changing patterns of demand

Patterns of agricultural demand are in flux. Rice used to be grown in poorly drained central parts of the basin but has been replaced by cotton, and there has been a steady increase in grape and fruit tree areas as agroindustrial enterprises have grown up to support cash crop agriculture. The trend toward grape cultivation, partly a response to the growing market for raisins, resulted in decreased demand for irrigation water, and total irrigation deliveries are now only about 70% of the pre-drought situation. With a recent surge in interest in drip irrigation by fruit, vegetable and seed maize growers, demand is likely to continue to decline.

In contrast, non-agricultural demand is growing rapidly. The area has a higher than average growth rate because of in-migration from poorer parts of Turkey, and Izmir's promotion of industrial development to complement agricultural production. Domestic demand for water has been growing by 2–3% a year and industrial demand by as much as 10% per year. Given that most non-agricultural consumption of water is from groundwater rather than surface water, aquifer management requires closer attention than surface water with respect to available volumes.

However, an additional demand is arising which is associated with growing concern over water quality, particularly during the peak of the summer season when surface water supplies are limited. Figure 11.3 shows actual and estimated growth of basin population between 1970 and 2010, along with estimated organic load from both domestic and industrial sources. As seen, although domestic load increases modestly along with population growth, industrial load grows exponentially. Note that the chart shows only potential loads created and does not take into account the effect of treatment facilities that may subsequently be built.

11.3 Legal, Policy and Institutional Environment

11.3.1 Water rights

All natural water resources, except some small privately owned springs, are vested in the State by the Turkish Constitution (Yavuz and Cakmak, 1996). The basic principle governing surface water use rights in Turkey provides that water is a public good which everyone is entitled to use, subject to the rights of prior users. Surface water use

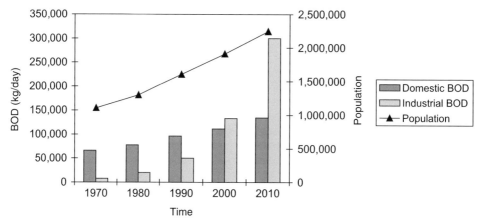

Fig. 11.3. Estimated Gediz basin population and organic waste load generation, 1970–2010.

is normally free of any obligation to obtain prior authorization. Conflicts are resolved by first referring to local customary rules and regulations. If the dispute cannot be resolved in this way, rights are settled by court decision. There is no registration system for surface water rights or water use. In large basins where impacts of new diversions are diffuse, this system is generally unable to prevent or resolve conflicts between new and existing claims, and this is leading to serious problems of over-allocation in some basins (Svendsen and Nott, 2000).

Groundwater also is State property. Its management is governed by a 1960 law giving sole authority over the use and protection of groundwater to DSI. Drilling a well deeper than 10 m requires prior approval from DSI, while constructing shallower wells requires only that DSI be notified. Shallow groundwater is thus an open access resource, while deeper aquifers are subject to some controls.

According to the groundwater law, when abstractions 'approach the safe output level' of the aquifer, a committee of representatives of 'relevant ministries' is to be formed to decide on pending and future applications for groundwater utilization. Frederiksen and Vissia (1998) conclude that enforcement of the groundwater law is weak, implying that, in practice, the system of groundwater rights created by the current groundwater law is not as effective as it could be.

Rights to both ground and surface water use are thus not formalized. Although they follow roughly the appropriative doctrine of allocation, there are no guarantees of continued access. The principles of the system of water rights outlined above apply in the Gediz basin.

11.3.2 Actors

In an earlier section, five categories of water users in the Gediz basin were identified and their respective water uses outlined. Some of these water users are able to represent their own interests (industries and municipalities), while others may be either many and disorganized (small system irrigators) or unable for other reasons to represent themselves (ecosystems). In addition, there are other State actors such as DSI involved in Gediz basin water management, which, while not water users, are important players. The range of basin stakeholders is thus different, and broader, than the group of actual water users. The major ones of these are described below.

11.3.2.1 Public agencies

DSI The DSI is the main executive agency of the Government of Turkey for the country's overall water resource planning, execution and operation. It was established in 1954 and is currently a part of the Ministry of Energy and Natural Resources. The mandate of DSI is 'to develop water and land resources in Turkey' (DSI, 1995). It is responsible for major irrigation, flood control, drainage, hydropower development and supplying water to cities with a population over 100,000. DSI centralizes most of the State functions involved in planning and developing large-scale water resources.

Until recently, DSI's policy has been to manage the irrigation schemes it designs and constructs. Current policy and practice is to transfer schemes to locally based IAs to manage. DSI also transfers hydropower and municipal water supply schemes that it designs and constructs to other agencies to operate.

DSI is also responsible for managing and allocating groundwater to prospective users. It does this through the permitting system described in the previous section. Its responsibilities for groundwater quality are limited to monitoring.

DSI maintains 26 regional offices across the country, organized along watershed lines. The Gediz basin lies entirely within one of these regions and is serviced by the regional office in Izmir.

MINISTRY OF ENVIRONMENT The Ministry of Environment (MoE) is the public agency with overall responsible for surface water quality. In spite of this general status, however, its mandate and capacities cover only some of the functions that implementing this responsibility entail. Its major responsibilities include coordinating plans among the various public and private agencies involved with protecting the environment, commissioning environmental impact assessments of major water resources projects, and setting standards for and monitoring surface water quality. Actual monitoring and reporting of water quality and wastewater discharges are carried out by provincial offices of MoE. The explicit mandate of the MoE does not extend to groundwater quality. Neither the national nor the provincial offices of MoE possess direct enforcement powers.

MUNICIPALITIES AND VILLAGES   Towns and villages play three important roles in the water resource arena. First, they are water users and dischargers of wastewater. There are 19 settlements in the basin with a combined population of 1.35 million. All draw their domestic water supplies from groundwater. Of the 19, only three have completed wastewater sewage systems and treatment plants. The remainder discharge untreated wastewater into the Gediz and tributaries.

The second important role played by towns and villages is that of representing irrigation water users in their areas. They do this: (i) through their statutory dominance of the boards of large-scale IAs; (ii) as owners and operators of municipal irrigation wellfields; and (iii) as representatives of the interests of otherwise disorganized farmers irrigating from private wells or small surface water sources who make up parts of their constituencies.

The third role is that of environmental regulation. Municipal and village administrations are responsible for operating water and wastewater treatment plants within their jurisdictions and monitoring the quality of domestic water supplies.[3] They also have some authority to monitor industrial

wastewater discharges, although most are not active in this area.

PROVINCES AND DISTRICTS   Provincial and district governors, appointed by the Ministry of the Interior in Ankara, are the only authorities with the power to assess fines or issue and enforce prohibitions against violators of water quality regulations. All other actors, including MoE, MoH, DSI and municipalities, may only report cases that contradict laws for which they are responsible to the provincial or district governor. District governors must secure approval from the provincial governor before taking action. Provincial governors are thus singularly responsible for water quality enforcement proceeding.

STATE PLANNING ORGANIZATION   The State Planning Organization is an arm of the Prime Ministry that prepares a rolling 5-year investment plan for the nation. It is responsible for planning all public capital investment in the country, including investments for water resource development, wastewater treatment and environmental problem mitigation.

GENERAL DIRECTORATE OF RURAL SERVICES   The General Directorate of Rural Services (GDRS), a part of the Prime Ministry, is responsible for developing small-scale groundwater resources for irrigation, developing surface water sources with flows of less than 500 l/s for irrigation, on-farm irrigation development, and the construction of rural roads and village water supply systems. GDRS's minor irrigation schemes are transferred to farmers' cooperatives or local governments upon completion. GDRS does not have an operation and maintenance capacity.

11.3.2.2 Semi-public or private groups

IAS   Thirteen Irrigation Associations (IAs) were established in the seven large canal

---

[3]   The Ministry of Health (MoH) may monitor the quality of drinking water in piped distribution systems at the request of a municipality or village.

irrigation commands in the basin in 1995 under the accelerated irrigation management transfer programme of DSI and have assumed operational control of canal irrigation in those areas. DSI continues to operate the main reservoir and river diversion structures, but operational management below that level is now in the hands of the IAs.

The legal basis for forming IAs is a law allowing the establishment of associations of local governments, and the present governance structure of the IAs is dominated by elected village headpersons, town mayors and elected members of local municipal councils. IAs are public bodies that enjoy tax exemptions and are non-profit, but are not bound by standard government civil service regulations and financial procedures. Although this system has drawbacks, it does provide valuable links with local government structures. IAs operate the canal systems within their areas, employing hired staff and financing operations and maintenance through fees collected from water users. The 13 IAs collaborate extensively on an informal basis and have discussed the possibility of forming a more permanent association to represent their common interests. They are the most important water users in the basin and retain a strong functional tie to DSI, which provides their bulk water supply and serves as the basin water allocation authority in the absence of a more explicit system of water right allocation.

OTHER IRRIGATORS   Other irrigation water suppliers and users not encompassed by IAs include towns and villages that have developed well-fields for irrigation supply in their areas, individual farmers and groups of farmers who have invested in irrigation wells, and farmers who employ small surface water diversions in upper parts of the Gediz catchment to irrigate crops. There is no formal organization tying these water users together, though their number is significant. To some extent, local village heads and town mayors are able and generally willing to represent the interests of these

irrigators when a need arises. Such representation is not coordinated among villages, however, and in general would not be expected to be particularly potent in competition with larger, better organized interests.

ENVIRONMENTAL NON-GOVERNMENTAL ORGANIZATIONS (NGOS)   There are many NGOs active in the field of environmental conservation in Turkey. A 1995 directory lists 98 of them, and there are others not included in the directory. With respect to water-related issues in general, and the Gediz in particular, the following are among the most important:

• Turkish Erosion Control, Reforestation, and Environment Foundation (TEMA). TEMA was established in 1992 with strong business community support. It currently has about 50,000 members and in 1997 operated on a budget of US$2 million. TEMA publishes a monthly bulletin on environmental issues and every 2 years publishes an *Environmental Profile of Turkey*, which is also available in English. It enjoys good contacts with the MoE, and has been instrumental in shaping the new national environmental laws and regulations.[4] It is the most influential of the national environmental NGOs.

• Gediz Basin Erosion Control Reforestation and Environment Foundation (GEMA). This NGO has interests similar to those of TEMA but is concerned specifically with the Gediz basin.

• Society for the Protection of Wildlife. This society was established in 1975 and works to raise awareness of shrinking populations of various wildlife species, with a special focus on birds. The society works extensively with elementary school children, publishing a newsletter and guidebooks for schools and others. It collaborates with the World-Wide Fund for Nature and other international organizations.

---

[4]   DSI has an agreement with TEMA for reforestation of certain catchment areas above DSI reservoirs.

Although concerned with water, none of these organizations places a priority focus on it. Most NGO activities to date have been concerned with education, awareness raising and lobbying, with little independent scientific or information collection effort evidenced.

INDUSTRIES   Although industrial plants are scattered throughout the Gediz basin, the largest concentration is in two Organized Industrial Districts, one in Kemalpasa in Izmir Province with about 180 enterprises, and the second near Manisa in Manisa Province with about 50 enterprises. The owners of these industries are organized into several associations, which wield considerable political power. These include the Aegean Chamber of Industry, the Businessmen's Association and the Young Businessmen's Association.

ENVIRONMENTAL ASSEMBLIES   Local environmental assemblies (Mahalli Cevre Kurulu) have recently been formed in several basin areas. Authorized under the Environmental Law, assemblies are broadly constituted, comprising mayors, DSI, the Chamber of Industry, and so on, and are typically chaired by the provincial governor. They meet monthly and are authorized to make fairly influential decisions on issues relating to urban environmental quality. Such an authority, chaired by the district governor, exists also for the Kemalpasa Organized Industrial District. A similar association was established for the Gediz delta bird reserve in 2002.

WATER AND THE POOR   As always, the poor, particularly those living in makeshift and illegal housing on the urban fringe, have the worst access to safe drinking water and sanitation services. Because of more limited mobility, they are also the ones most affected by pollution of the Gediz, which they are more likely to use for recreational purposes. Access to irrigation water is determined by access to land, which in turn is also related to wealth, both as a cause and as an effect. Many smallholders do practise very productive agriculture, however, often growing high-value horticultural crops.

11.3.3 Essential functions: gaps and overlap

Essential functions of basin management are described and defined in Chapter 1. An essential functions matrix for the Gediz basin, crossing functions with key actors, is shown in Table 11.2. These functions are replicated, as appropriate, across four broad categories – surface water, groundwater, wastewater disposal and agricultural return flows. Cells are marked to indicate an actor active in a particular functional area. Information is drawn largely from richly detailed reports prepared by Harmancıoğlu and Alpaslan (1999, 2000a,b).[5]

A number of interesting points emerge from an examination of Table 11.2, supplemented with background observations:

• There has been very limited planning at the basin level with respect to surface water and virtually none for groundwater and waste disposal.[6] There is no integrated plan that considers both ground and surface water availability, nor does existing planning consider water quality, wastewater disposal, current and projected land use, anticipated future demand and return flows, or projected future quantity and quality of water resources.

• Water is allocated, in practice, by a variety of agencies and users operating independently of each other. These include DSI, private surface and groundwater irrigators, municipalities and industries. There is no national

---

[5]   Note that the activity indications contained in the table refer to actual activity in practice, and not nominal responsibility as assigned in statutes. Open circles indicate limited activity, while filled circles indicate more extensive activity. Situations where there is only minor activity might not be indicated in the table. The indications are the collective judgements of the study authors and do not represent formal or official judgements by any of the collaborating organizations.

Table 11.2. Key actors and essential basin management functions in the Gediz basin.

Key actors	Surface water									Groundwater							Wastewater						Ag. returns			
	Plan (basin-level)	Allocate water[a]	Construct facilities	Distribute water	Maintain facilities	Monitor quality	Ensure quality	Protect against flooding	Protect ecology	Plan (basin-level)	Allocate water	Construct facilities	Withdraw/distribute	Maintain facilities	Monitor quality	Ensure quality	Plan (basin-level)	Construct facilities	Operate facilities	Maintain facilities	Monitor quality	Enforce quality	Construct facilities	Maintain facilities	Monitor quality	Enforce quality
DSI	○	●	●		●	●		●	○	○	●	●			●								●	○		
Irrigation associations			○	●	●																			○		
Other irrigators		○	○	●	●						○	●	●	●												
GDRS			●									○											○			
MoE						○			○												○					
Local governments												●	●	●				○	●	●						
Industries											○	●	●	●				●	●	●						
Provinces (MoI)																						○				
NGOs																										
Bank of the Provinces (IB)																		●								

●, Indicates activity; ○, indicates limited activity.
DSI, State Hydraulics Works Organization; GDRS, General Directorate for Rural Services; MoE, Ministry of Environment; local governments, locally elected urban governments (municipalities and villages); MoI, Ministry of Interior; NGO, (environmental) non-governmental organizations.
Note: Surface water is used only for irrigation and environmental purposes.

legal framework for surface water rights and only a rudimentary system of allocating access to groundwater, and, as a result, both are largely open-access resources. Although nominal control is stronger for groundwater than for surface water, groundwater is presently the most stressed of the two resources. It is also the more desirable of the two, in part because of the poor quality of Gediz surface water in the lower reaches, but also because of the relative ease of access provided by groundwater. The current system of registering groundwater withdrawals does not appear to be effective at limiting overdrafting, which is occurring in certain sub-basins.

- Water quality monitoring takes place, but information is often not available in useful forms to interested parties. DSI operates 14 water-quality sampling locations within the Gediz basin, sampling about 35 parameters on a monthly or bimonthly basis. The information collected, however, remains as raw data in DSI files and is not generally used as a basis for policy making or decision making for basin management.
- A single actor, the provincial governor's office, is empowered to authorize enforcement of breaches of wastewater discharge regulations by banning offending practices or imposing fines. In practice, attempts to process fines or prohibit industrial activities often leads to confrontation between industrialists and public administrators, with the administrators generally lacking the political will and power to make penalties stick. It is very common for files of violation reports to remain unprocessed in the offices of district and provincial governors.

- Ensuring surface and groundwater quality is not actively practised in the Gediz basin. Ensuring water quality involves conducting follow-up investigations of observed substandard water quality to identify its sources and proposing remedies.
- No attention is currently paid to the quality of agricultural return flows. It is sometimes presumed that these flows contribute nitrates to groundwater and nitrates, phosphorus and organic chemical residues, e.g. from pesticides, to surface water, but there is little hard information on this, nor is any responsible party actively monitoring or assessing the quality and impact of agricultural return flows.
- Agricultural drainage infrastructure is inadequately maintained at present. Drain maintenance receives a lower priority from IAs compared with delivery channel maintenance and DSI does not have an adequate budget and equipment to fully maintain larger drains. In addition, responsibility for maintaining main drains, which serve more than one IA, is under dispute by DSI and IAs.
- NGOs have no specific role in performing essential management functions, but clearly have an important role to play in overall basin governance.[7] This leads to the discussion of enabling conditions in the following section.

11.3.4 Enabling conditions: where problems lie

The essential functions and actors' roles depicted in Table 11.2 provide a static view of responsibilities. Additional attributes of well-functioning basin governance systems

---

[6] DSI is currently anticipating a new Gediz basin planning exercise. The previous plan was prepared 35 years ago and updated in 1982, but covered only surface water. The groundwater section of the DSI regional office is also planning a new groundwater survey in the near future.

[7] The term *governance* is used refering to the rules providing the context for multi-actor basin management and the processes and activities engaged in by those actors operating within this set of rules.

relate to its dynamics. These attributes, which we term *enabling conditions*, were defined in Chapter 1 and are shown in Box 1.2. A full analysis of these factors is well beyond the scope of this paper. A brief sketch of each in the context of the Gediz basin will be attempted to illustrate the concepts and indicate broad strengths and weaknesses.

11.3.4.1 Political attributes

This is perhaps the most important gap in the current set of enabling conditions. Although some water users are well represented, others are not, and in the arena of political give and take, those without representation become losers. *Industrialists*, for example, have ample financial resources, are well organized and have ready access to political decision makers. *Other irrigators*, on the other hand, are not organized and enjoy representation only through their local village heads. *IAs* are intermediate. They enjoy multiple connections to the local political establishment by virtue of having a number of village heads and town mayors on their managing committees. In addition, they collaborate informally, sharing information and coordinating activities. IAs would benefit by establishing more formal linkages among themselves to allow a single spokesperson to represent them collectively in discussions over basin water allocation, water quality standards, irrigation return flow restrictions, and so on. Other irrigators could affiliate with such an association and, contributing financially to it, participate in its representational benefits.

The most serious failure of representation currently relates to the environment. Although nominally represented by the MoE, the Ministry is still relatively young and has yet to establish presence and capacity in many areas. For example, it currently has provincial offices in only two of the four provinces covered by the Gediz basin. It also lacks sufficient budget to perform its many duties fully. Moreover, as a government agency, it will always be subject to political pressures and pulls that encourage or inhibit vigorous pursuit of particular water quality issues. Experience from other countries has shown that strong NGOs rooted in civil society are essential components of the political system surrounding environmental issues. These NGOs serve as advocates for environmental values and for unrepresented future generations. There are several groups with potential to fulfil this role, but they presently provide an ineffective counterweight to other interests.

Just as important as the existence of representational bodies is the need for a rough balance of political power and influence among various interests. When power is one-sided, issues are not aired adequately, and decisions are also one-sided. A key to the evolution of a suitable and balanced governance regime for the Gediz basin is further maturation of non-government organizations and associations based in civil society which can advocate for environmental interests.

11.3.4.2 Informational attributes

Stakeholders in the basin need to have access to accurate and up-to-date descriptive information on water-related issues in the basin, and open access to decision-making processes related to plans, regulations, violations and sanctions to participate effectively in basin governance. Data collection in the Gediz basin is incomplete, in that it falls short of the set of data required for competent professional management of the basin's water resources. Data that are collected, however, are often not widely disseminated or, worse, are sequestered and jealously guarded by the collectors. This retentiveness may stem from feelings that the data are a source of power if they are not widely available, or perhaps from doubts about the reliability of the data. In a larger sense, though, it reflects the lack of a strong demand for the data for use in basin planning and management. Once active planning and management processes are mounted, the pressures for access to data will grow and push much of this data into the open.

11.3.4.3 Legal authority

Establishing appropriate institutions requires suitable legal authority. This does not appear to be the most serious current problem constraining the emergence of an effective governance regime in the Gediz basin. Although improvements can be made, and a number of legislative changes are proposed in the Eighth Five-Year Plan, the most important short-term constraints appear to lie in other areas, such as balanced political power and providing adequate resources. Over the longer run, however, the legislation of a new legal basis for an effective system of water rights allocation, protection and transfer will be essential.

11.3.4.4 Resources

Clearly, all four types of resources listed in Box 1.2 are needed for effective implementation of basin management activities. In a number of the responsible public organizations, they are inadequate. In some, this constraint may be relaxed by reassigning staff positions from functions that have lost importance to those that are increasingly so. Another potential problem is scattering of resources among a variety of institutions, where each lacks a critical mass to be effective. In a context of cooperation, it is not necessary that resources be consolidated under a single administrative structure for effective implementation. However, cooperation and coordination must be effective if a decentralized strategy is to be effective.

11.4 Toward Solutions

11.4.1 Problem recap

In this section, the hydrologic, policy and institutional problems identified in earlier sections are brought together and summarized.

11.4.1.1 Poor surface water quality

The most pressing water-related problem facing the Gediz basin presently is the poor and deteriorating quality of its surface water. The deterioration results primarily from the basin's recent rapid growth in population and the even more rapid growth in local industry, coupled with the widespread use of agricultural chemicals in a highly productive agriculture. Failure to control this growing problem at its several sources leads to large requirements for in-stream flows for dilution – flows that are then unavailable for other uses. The problem stems from several sources:

- *Weak enforcement* – first and foremost, it is the inability of the provincial and district governors, appointed by the Ministry of Interior, to apply and enforce sanctions and penalties on violators of wastewater discharge standards that is responsible for the growing pollution problem. Although monitoring could be improved and standards tightened, the failure to effectively enforce existing standards, based on existing information, sends a powerful signal to polluters that compliance is unnecessary.
- *Weak coordination* – a second cause of deteriorating surface water quality is poor coordination and cooperation among the three separate agencies responsible for: (i) surface water quality monitoring; (ii) wastewater discharge monitoring; and (iii) enforcement of standards. To some extent, this is driven by bureaucratic tussling over turf. In addition, the failure of any of the three parties to come forward with effective, inclusive and forward-looking leadership is a cause.
- *Limited availability of data* – because of restricted access, the debate on water quality is poorly informed and emotional rather than scientific, making development of acceptable remedial measures difficult and contentious.
- *Haphazard monitoring of wastewater discharges* – the most readily identifiable and correctible causes of Gediz pollution are untreated or inadequately treated wastewater discharges from

industries, cities and towns. It is the responsibility of the MoE to monitor these discharges and report breaches of standards to the provincial governor. Limited staff, laboratory facilities and funds currently forestall an adequate monitoring programme.

- *Inadequate funding for wastewater treatment plants* – inadequate funding has two components, capital and operating expenses. Several funding windows are available to industries and municipalities for investment capital but a shortage remains. For municipalities, there is little private sector involvement in constructing and operating treatment facilities, in contrast to the case in many other countries. A considerable share of the funds made available by the State for municipal treatment plant construction have come through the Bank of the Provinces, but with little expectation of repayment. This makes the discipline of mobilizing private capital for such investment difficult or impossible. Inadequate investment in wastewater treatment by the industrial sector relates, in part, to the weak enforcement record of provincial governors. As long as the costs of compliance exceed the costs of non-compliance, this situation is likely to continue.

- *Limited public awareness of the problem* – negative effects of surface water pollution include harm to public health, increased costs to other water users and negative environmental effects, particularly in the Gediz delta. Limited public awareness of the problem and its impacts results in limited public pressure and support for reforms, which in turn affects every single one of the factors outlined above. Causes of limited public awareness include restricted public access to water quality data collected by government agencies, inadequate MoE efforts to publicize water quality problems, and the failure, to date, of environmental NGOs to become effective advocates and spokespersons.

11.4.1.2 Unknown groundwater quality

The extent of possible groundwater contamination, particularly in the Nef Creek watershed, is unknown. Due to coarse alluvial soils and the extensive use of in-ground holding pits, some wastewater from both urban and industrial sources may go into groundwater rather than being disposed of as surface water effluent. Groundwater quality monitoring is not widespread and the results not publicly available, making it difficult to know if significant degradation of groundwater quality is occurring.

This is a significant gap because aquifer pollution is often more difficult and expensive to mitigate than pollution of surface waters. The possibility of contamination gains added significance as a result of the almost total dependence of the basin's population on groundwater for domestic supplies.

11.4.1.3 Loosely controlled allocation among users

Shallow groundwater is an open access resource in the Gediz basin, meaning that anyone with physical access to such water can withdraw and use it. Deep groundwater and surface water, once released from Demirkopru Reservoir, share this open access characteristic, in part, as well. The result is that some legitimate needs, especially environmental needs, are inadequately met, access of existing users is insecure and it is difficult to transfer water allocations among users in a rational way. Among the causes are the following:

- *Inadequate representation* – interests of some users are not well represented in allocational planning and decision making. The most salient example is the environment, and, in particular, the needs of the Gediz delta and its rich complement of wildlife.

- *Inadequate specification* – while water needs in the basin for large-scale irrigation and urban use are generally known, present use and future requirements for small-scale irrigation, the burgeoning industrial sector, and the environment are not well specified in terms of quantity, timing and quality requirements. This makes allocation decision making difficult.

- *Ineffective reporting and record keeping* – while municipal extractions are reasonably well documented, industrial extractions from groundwater remain largely a matter of conjecture. This makes evaluation of new requests for withdrawal permits difficult. There is likewise no cumulative inventory of the total number of permits issued, the agreed extraction rates and the depth from which water is extracted, rendering this process even less rigorous and reliable.

11.4.1.4 Overarching future problems

In addition to the current problems affecting water allocation and basin governance, there are longer-term problems that will require more fundamental changes in laws, policies, institutions and practices. Two of the most significant are the following:

- *Rudimentary water rights system* – the current national system of recording and harmonizing rights to use water dates from an earlier simpler time and is not well adapted to a water-short environment. It does not provide security for present users, does not allow for or adequately protect environmental uses of water, and does not provide incentives for economy of use or for orderly transfers among sectors.
- *Lack of integrated planning* – assessment of basin water resources is currently separated into ground and surface water components. Because these interact in practice, the basin needs to be understood as an integrated water resource system. Also, because water quality influences the uses to which water can be put, and gives rise to its own quantitative demands

for dilution flows, quantity and quality must also receive joint consideration.

11.4.2 Recent strides

The current situation in the Gediz basin, and at the national level, is dynamic and somewhat fluid. Locally, a number of steps have been or are being taken to improve the enforcement of water quality standards and protect the natural environment. Many of these steps began with a Franco-Turkish basin study of the Gediz in the mid-1990s. Although this study was intended to lead to various action programmes, the latter failed to materialize because of lack of co-operation among the different institutions involved.[8] The study did raise awareness of problems and stimulate other initiatives, however.

In 1996, the provincial MoE office conducted a study of polluting industries in Izmir province, which resulted in sanctions on 14 firms. In that same year, MoE offices in Izmir and Manisa began a 2-year surface water-quality sampling exercise in the Lower Gediz.

In early 1998, three provincial governors from the Gediz basin, together with MoE and other parties, convened a 'coordination meeting' for the basin, which led to the establishment of a coordinating committee consisting of the directors of the three provincial MoE offices. More recently, this coordinating committee was transformed into a permanent body named the *Environmental Protection Service Association of Gediz Basin Provinces*. This Association was officially authorized by the cabinet of the national government in December 1999, giving it a legal persona. This association has a broader base than earlier initiatives, and, in principle, has considerable power. Since its creation, however, it has lain largely dormant due to lack of resources and the ongoing bottleneck in the enforcement of existing standards and regulations.

---

[8] The study was never formally 'accepted' by DSI or MoE.

At the national level, sentiment for change is reflected in the recently published Eighth Five-Year Plan covering the period 2001–2005. The Plan recognizes the need for change in the way water is allocated and managed. There is a commitment to introduce new water legislation that will cover such issues as water rights, responsibilities for water allocation, setting and enforcing environmental standards, and consolidating the position of the IAs. However, the Plan does not specifically mention basin-level water management nor is there any provision for establishing basin-level entities that could implement or coordination various basin-level activities. Reduction in support for utilization of fertilizers and agricultural chemicals is a preliminary step toward addressing non-point source pollution problems. Revising water and wastewater standards to comply with EU standards will raise the bar for existing polluters and those in compliance alike.

The Plan does indicate that the private sector will become more involved in various aspects of water, adding an additional set of regulatory challenges to the existing situation and raising the question of the security of rights to water use by less well-represented groups. Until these new initiatives are defined and brought into place, however, current institutional arrangements will continue within the context of rapid growth in demand for water and increasing pressure on water resources from wastewater disposal from urban, industrial and agricultural users.

11.4.3 Strengths to build on

Although the problems faced are formidable, Turkey and the Gediz basin have a number of strengths on which to build an effective basin governance regime. These include the following:

- The premiere water resource agency in the country, DSI, is responsible for both ground and surface water, providing a strong base for integrated treatment in the future. This is not the case in many other countries, where separate organizations are responsible. Moreover, handing over irrigation management responsibilities to IAs positions it well to take on the role of basin planner and water quality monitor for both ground and surface waters.

- Water quality is squarely identified as an important problem in the Gediz basin. Moreover, while serious, it has not yet reached catastrophic proportions, offering a grace period in which action can be taken. Some actors in the basin appear to be responding to the warning signals.

- There is recognition that a number of different actors must be involved in solving water quality problems in the Gediz. It is important to transform this recognition into effective ways of working together, rather than squandering energy and resources in intra-governmental squabbles over bureaucratic turf.

- Likewise, there is recognition that there are multiple dimensions to water resource management problems – different disciplines, different interests, different uses, ground and surface water, quantity and quality, and so on. Recognizing this provides opportunity to develop an integrated approach to basin water resource planning and management.

- A new water law is under consideration, offering an opportunity to lay legal groundwork for effective basin management and protection for the Gediz and other water-short basins in the country.

- There is a strong university-based scientific community, e.g. CEVMER and others in or near the basin, providing capability for applied problem-solving research and, where needed, independent scientific assessment.

- There are linkages with international institutions such as IWMI, which provide access to international experience with basin governance problems. Moreover, there is a healthy willingness to look outside the country to the experiences of others and a strong interest in harmonizing standards, practices and procedures with those of the EU.

11.4.4 Challenges

A number of immediate solutions to problems affecting governance and management of water resources in the Gediz basin are self-evident from the listing of problems in Section 11.4.1. In this concluding section, we indicate four important longer-range challenges facing the basin. Addressing these challenges effectively would go a long way toward putting into place a strong, dynamic and flexible system of basin governance.

11.4.4.1 Systematize water rights

The current rudimentary rights system cannot provide security and flexibility in an era of growing water scarcity. A new system which is fair, flexible, and effective needs to be designed, based on both Turkish and international experience.

11.4.4.2 Build representational presence and political muscle

Some basin water users, such as the natural environment, are not well represented in water-related discussions at present, and there are severe imbalances of political power among the various water users. Fair and equitable governance of basin water resources requires that users and interests be represented in discussion and decision-making fora in a balanced way. NGOs rooted in civil society provide an important voice and advocacy presence for the environment, supplementing the efforts of the MoE. Their emergence as a political force will add balance to decision making and

pressure for effective enforcement of sanctions for water quality violations.

11.4.4.3 Develop coordinating mechanisms

Alternative models for water basin governance exist. At one pole is a *comprehensive basin authority*, which concentrates power, responsibility and capacity to implement directly many basin management tasks. At the other pole is a *coordinating committee*, which simply provides a forum for discussion and voluntary coordination. Between these poles, many variations are possible. One thing that is clear is that the present system of compartmentalizing water quantity and quality, ground and surface water, and fresh water and wastewater is not an effective base for the future. Mechanisms have to be developed for bringing these components together in a functional integrated system for planning, governance and management.[9]

A useful first step would be the completion of an integrated assessment of basin water resources of all types and of the present and predicted demands on those resources. High-level political commitment to such an undertaking and strong leadership would be essential. It is equally important that this exercise not be carried out by a single organization, but that it involves the various agencies, and the different departments within agencies, which have mandates to address water-related issues in the basin. The report produced is only half of the desired output of such a process. The other half is the experience of joint action among agencies and groups to implement the study and the creation of 'ownership' of the result by the various stakeholders.

11.4.4.4 Involve the private sector

In many countries, the private sector plays important roles in water resource management. Turkey is well embarked on the

---

[9]   One reason that Turkey is not pursuing the concept of basin-wide authorities at present is because of the difficult issues posed by important trans-national river basins shared by Turkey and several of its neighbours. It is said that the next 5-year plan may address this option.

devolution of responsibility for managing previously State-operated irrigation systems to locally based associations. The private sector can also play a major role in providing safe drinking water and effective sanitation services to urban areas. Private sector involvement has a number of advantages, including operational efficiency, ability to mobilize private capital and access to new technology. To attract such involvement without State guarantees of repayment, firms must have confidence that the principle of payment-for-service will be honoured and supported by the involved governmental entities. Bringing in such private participation would provide needed capital for wastewater collection and treatment systems and provide wider access to these essential services.

References

DSI (1995) *DSI in brief*. Ankara.

Fredericksen, H. and Vissia, R. (1998) *Considerations in Formulating the Transfer of Services in the Water Sector*. IWMI, Colombo, Sri Lanka.

Harmancıoğlu, N. and Alpaslan, M.N. (1999) *Institutional Support Systems Project, The Turkey Activity, Interim Report I*. CEVMER, Dokuz Eylul University, Izmir.

Harmancıoğlu, N. and Alpaslan, M.N. (2000a) *Institutional Support Systems Project, The Turkey Activity, Interim Report II*. CEVMER, Dokuz Eylul University, Izmir.

Harmancıoğlu, N. and Alpaslan, M.N. (2000b) *Institutional Support Systems Project, The Turkey Activity, Final Report*. CEVMER, Dokuz Eylul University, Izmir.

Kayam, Y. and Svendsen, M. (1999) *Small Scale Irrigation in the Gediz Basin*. Report prepared for the IWMI/GDRS joint research project on Irrigation in the Basin Context. IWMI, Colombo, Sri Lanka.

Svendsen, M. and Nott, G. (2000) Irrigation management transfer in Turkey: process and outcomes. In: Groenfeldt, D. and Svendsen, M. (eds) *Case Studies in Participatory Irrigation Management*. World Bank Institute, Washington, DC. *Wall Street Journal*, 4 May.

Yavuz, H. and Cekmak, E. (1996) *Water Policy Reform in Turkey*. Report, DSI and Istanbul University.

12 Managing River Basins: Lessons from Experience

Mark Svendsen and Philippus Wester

12.1 Introduction

The case studies presented in the preceding chapters provide a rich vein in which to prospect for common threads and meanings. The six cases vary along a number of dimensions and, at the same time, each is, in some respects, unique. Moreover, the case studies employ similar, though not identical, formats. This chapter will compare the cases and attempt to draw conclusions that will aid those establishing and strengthening basin management capabilities elsewhere.

We first ask what drives countries to establish and maintain mechanisms for basin coordination and management. Next, we describe the mechanisms that have emerged in the six case study basins, contrasting selected characteristics. Then we look at the conditions which enabled these mechanisms to emerge, or which prevented them from emerging. Finally, we summarize lessons learned.

12.2 What Drives Basin Management?

In a matter as complex as the emergence of human institutions, it would be astonishing if a single variable explained the emergence conclusively. In this case, we are not to be astonished, as a number of factors appear to come into play, while other plausible influences do not seem to have a strong effect. If the sample of cases were wider, no doubt other factors would rise to the surface as well. Table 12.1 shows the six cases arrayed against a number of dimensions related to the possible emergence of basin management mechanisms.

At this point, we make the summary judgement that the first three basins – those in France, California and Mexico – have put in place mechanisms to manage water at the basin level, while the other three have not yet done so. The mechanisms in the first three basins are very different from each other and operate with different degrees of effectiveness, but they do exist and operate. In the study basins in South Africa and Vietnam, governments intend to develop such mechanisms and have taken preliminary steps in this direction but neither is yet operating. In Turkey, there are as yet no clear plans to introduce formal basin management practices. We will examine the nature of these mechanisms subsequently, but for the moment it is interesting to explore some of the possible driving forces relative to this simple 'with' and 'without' dichotomy.

The first dimension, the size of the basin in question, seems to have little or no effect on its propensity to develop management mechanisms. It might be supposed that larger basins would have a stronger need for organized management. Conversely, it could be thought that smaller basins would be easier to manage and therefore more likely to

Table 12.1. Selected basin characteristics.

Basin	Country	Watershed area (km²)	National PC income[a]	Quantity constraint	Quality problem	Groundwater overdraft[b]	Downstream environmental assets[c]	Storage[d]
Neste	France	10,000	$24,080	●		–	?	< 40%
Central Valley	California (USA)	158,000	$34,280	●		112%	●	Substantial
Lerma-Chapala	Mexico	54,000	$8,240	●	○	116%	●	78%
Olifants	South Africa	74,000	$10,910	○		–	●	105%
Dong Nai	Vietnam	40,000	$2,070			–	○	13%
Gediz	Turkey	17,000	$5,830	○	●	137%	●	128%

●, Major; ○, minor.
[a]PC, per capita income; 2001 PPP based on World Bank World Development Indicators.
[b]Withdrawal as a percentage of average annual recharge.
[c]Sensitive environmental area at downstream end of the basin.
[d]Controllable storage as a percentage of average annual discharge.

develop a management system. However, both the smallest watershed and the largest in the sample fall into the 'with management' category. Moreover, the two smallest watersheds – those in France and Turkey – clearly are of great importance in their respective countries, and yet one has a very sophisticated management set-up while the other has virtually none.

Per capita income, which would indicate the ability of an economy to pay for basin management services, is a potential enabler of formal basin management. In this case, the sample of case studies offers some support for this hypothesis, since the two wealthiest countries clearly have the most sophisticated basin management set-ups. However, the strong steps being taken by Mexico contrast sharply with the virtual absence of action on the part of Turkey, though both countries have similar income levels, and if regional income measures were compared, the Izmir/Gediz area would doubtless show a considerably higher per capita income than both the Turkey and Mexico averages.

A second connection between higher levels of per capita income and basin development is that per capita income serves as a rough proxy for a higher level of institutional specialization and diversification. This has implications for a variety of things, from data collection and analysis capacity to the availability of a large pool of trained enterprise managers. It is likely that this effect is also operating here, though lower national incomes are certainly not an absolute constraint on the emergence of basin management mechanisms.

It is probably significant that all three of the 'with management' basins are experiencing severe surface water availability constraints, while in the three 'without' basins the constraint is less severe. The judgement made here on 'quantity constraint' is not as rigorous as is the designation of a 'closed basin' in the depletion methodology outlined in Chapter 3. In particular, the 'closed basin' definition effectively requires that the temporal variability in water supply has already been smoothed out through the creation of in-basin storage, since this is the

heart of the stair-stepped water availability function in that concept. In this sample of cases, the Lerma–Chapala has storage for nearly four-fifths of the annual runoff of the basin, and the construction of additional storage in both the Neste and in California is severely constrained by public policy and environmental considerations. It is thus reasonable to regard these basins as essentially closed. The Oliphants and Gediz basins both have the capacity to store more than the annual basin runoff and have utilized most of the available supplies, but effective demand has not yet risen to the point where there is serious conflict over existing resources. As historical inequities are redressed in the Olifants, however, demand will doubtless expand to exceed supply. In the Gediz, the agricultural demand for water is declining as cropping patterns shift under economic pressure to less water-intensive crops, reducing the pressure on the resource from this source. The Dong Nai is lightly used at present and has ample scope for additional storage to smooth out seasonal peaks in water supply. While seasonal constraints on availability are approaching in the Dong Nai, availability has not begun to approach the utilizable limit.

So in all three 'with' basins, serious competition over available basin water supplies appears to drive basin management efforts, supporting the fundamental premise developed in Chapter 2.

Water quality concerns have not, by themselves, been sufficiently strong to drive the creation of basin management mechanisms in the two basins where they are most prominent – Lerma–Chapala and Gediz. Similarly, groundwater overdrafting, which is serious in three of the basins, does not figure prominently in basin management plans, where they exist. In fact, in the Gediz basin where overdrafting is the most severe, basin management has received the least attention among the six cases.

Four of the basins have important environmental assets at their lower end. In California, the Sacramento–San Joaquin Delta is the focal point of environmental concern and controversy in the state, and in Mexico the level of Lake Chapala is seen as a primary indicator of the success of management efforts. In the Olifants the world renowned Kruger National Park lies astride the river on the South African border with Mozambique, and in the Gediz the estuary where the river enters the Mediterranean is an important bird sanctuary recognized internationally as a Ramsar site. These four sites are evenly divided between the 'with' and 'without' basins.

From this discussion we can hypothesize that absolute limits on available water supply in practice provide the strongest impetus for creating formal basin management mechanisms and that wealth and institutional resources in the larger society are important enabling factors for their emergence. Where water supply expansion is feasible, it is likely to be a preferred option to more intensive management and conservation of existing supplies, as seems to be the case in the Dong Nai. Water quality and environmental protection problems do not seem to be sufficient motives, by themselves, to cause the emergence of comprehensive management mechanisms. Equally, overdrafted groundwater, while worrisome, is not a strong driver of management efforts. In part, this may be because groundwater overdrafting is a devilishly difficult collective problem to get a grip on, and also because the problem is 'invisible' and its consequences lie primarily in the future.

12.3 Management Mechanisms

12.3.1 Configurations

12.3.1.1 Neste

The Neste is managed by a carefully structured set of interlinked organizations that cover all major basin management functions. It was first put into place about 40 years ago. Management is focused on surface water, since groundwater is unimportant in this basin. The centrepiece of the structure is a Regional Development Company (RDC) – a state-owned firm holding an operating concession for public water rights

and facilities that plays major roles in constructing and maintaining facilities and in allocating and distributing water to users. Water users comprise towns and municipalities, individual agricultural water users, and a few water users' associations (WUAs).

A Basin Committee or 'water parliament' made up of local and national authorities and water users develops long-term policies and plans for the basin. An executive arm of the Basin Committee, the Water Agency, collects water and pollution-based taxes and monitors water quality, but does not take a direct role in water management.

Regulatory oversight and 'policing powers' for water quality and environmental protection are provided by the Ministry of Environment. The Ministry of Agriculture, which represents the State as the owner of the infrastructure, plays a regulatory role in monitoring the care and maintenance given to the facilities by the concessionaire, the RDC.

The RDC is not involved in managing or regulating derivative water from municipalities and agriculture, which is handled by towns and individual farmers, supervised by the Water Authority and the Ministry of Environment.

The organizational set-up thus contains: (i) a broadly based representative body, the Basin Committee, supported by its executive arm, the Water Agency, to make plans and set policy directions; (ii) a government-owned company to build, operate and maintain hydraulic facilities; and (iii) public regulatory agencies comprising the Ministries of Environment and Agriculture. Notable by their relative absence are irrigation associations providing irrigation service to their members and representing their interests and other pressure groups representing environmental interests. The State, on the other hand, has a strong presence in all of the important organizations in the management set-up.

12.3.1.2 Central Valley

Institutional arrangements for managing California's Central Valley water resources stand in marked contrast to the planned and orderly set-up in the Neste. California's institutions have grown up by accretion over more than 100 years of resource development into a boisterous confluence of actors, laws, policies, plans and influence. Here two important categories of actors not figuring importantly in the Neste arrangements, interest groups and the courts, are involved, and in each of the major categories of involved actors – managers, service providers, users, regulators, advocacy groups, elected officials and the courts – there are multiple actors.

Major water resource development investments made in California over the past 100 years have blurred basin boundaries by making it possible to transport water across basins from north to south, and from the Colorado River basin west to southern California. This infrastructural development has multiplied the options available to managers, but at the same time, it has made traditional watershed boundaries somewhat obsolete as delimiters of responsibility and authority. Thus, while California has never had a basin management authority in the traditional sense, such an authority might well have become obsolete by now had one been established.

Instead of a formally structured authority to manage the waters of the Sacramento and San Joaquin Rivers, what controls is rather a body of laws, rules and plans that govern water allocation and reallocation, water quality and return flows, and which protect the natural ecology. Changes in existing practices and responses to changing conditions are contested by a variety of interest groups before decisions are reached. Federal and state courts usually serve as the final arbiters in these contestations.

The state Department of Water Resources (DWR) is the primary water resource manager, planning water development and use in the basin, constructing and maintaining major facilities, and protecting against floods. The United States Bureau of Reclamation (USBR) also constructs, operates and maintains major water control and supply facilities as does a large private consortium of irrigation water users.

The users of water in the Central Valley consist almost entirely of user-controlled water districts. These districts are non-profit, quasi-municipal entities governed by farmers, in the case of irrigation districts, and by urban residents in the case of municipal water districts. The districts receive water from the DWR or the USBR and, in turn, provide water services to their member farmers or households. These water districts have been features of the institutional landscape in the Central Valley for more than 100 years, and continue to operate reliably. They are linked together in associations and are thus influential actors in basin water management decision making.

Environmental protection and mitigation is a critically important issue in California and nearly half of the state's developed water resources is allocated to environmental purposes. Regulating all uses and protecting environmental resources are several powerful federal and state agencies. The important state Water Resources Control Board and the Regional Water Quality Control Boards are independent of the DWR and are insulated to some extent from other organs of state government as well. In addition, influential mass-based environmental action groups actively lobby for particular causes, join court cases, publicize and provide their own analyses of water/environment interactions.

Groundwater is far more lightly regulated than is surface water in the Central Valley and is seriously overdrafted. No state agency takes effective responsibility for regulating groundwater use and the legal basis for such regulation is deficient.

The regional water quality control boards enforce quality of return flows from both municipal uses and from agriculture rigorously, with support from the courts as needed. Municipal return flows are becoming increasingly important as water sources for agriculture, landscape and turf irrigation, aquifer recharge, and even new municipal supplies.

In sum, institutions are in place to construct, maintain and operate water facilities in the Central Valley, to protect water quality and the environment, to distribute water to end users efficiently, to reallocate water among uses using both market and administrative mechanisms, and to plan for future water supply and use. However, specialized basin-based management organizations do not exist and are unlikely to emerge. User-governed water supply organizations provide services to end users in most cases, and they remain healthy and effective. The institutional glue that holds the management system together consists of an accumulated body of legislation and case law, legally enforceable contracts, environmental rules and regulations, and an inclusive planning process led by the state DWR every 5 years. Market mechanisms play an increasingly important role in reallocating water to new and more highly valued uses.

12.3.1.3 Lerma–Chapala

Water resource management in Mexico operates under the strong centralized authority of the National Water Commission (CNA). CNA is a young organization, created in 1989, and operates under a comprehensive water law promulgated in 1992. It is empowered to manage national water resources, plan development and management, assign and register water rights, collect water-related taxes, and develop irrigation and urban water supply systems. It operates through a set of 13 regional (basin) and 31 state offices.

Day to day management of irrigation systems has been handed over to irrigation districts, similar to those in California, in a highly successful management transfer programme dating from the early 1990s.

In the Lerma–Chapala basin, a Basin Council was established under the new water law in 1993. A Governing Council chaired by the CNA Director General is the top level decision-making body, and a Monitoring and Evaluation Group (MEG) acts as its executive organ. The regional CNA office serves as a secretariat for the Council. A basin-level users assembly brings together the users of the basin and selects representatives from the various sectors to the Council. Council working groups address particular problems. Although the Basin Council provides a prototype vehicle to manage basin water resources, it has little decision-making power, since CNA remains responsible for major decisions, including water concessions, the collection of water taxes and water investment programmes.

The Lerma–Chapala Basin Council displays some of the features of the French Neste model, such as a users assembly, a top-level governing body, and strong State involvement and direction. At the same time, the set-up in Lerma–Chapala lacks the professional management expertise of the French Regional Development Company, its councils and assemblies have little of the decision-making authority of their French counterparts and they are heavily dependent on the CNA for authority, funds and technical expertise.

An analysis of actors and functions shows that, as in California, most of the essential functions related to surface water are covered, while groundwater is only weakly regulated. Planning, water allocation and water distribution are done reasonably effectively, while other functions are not. Demand management and reallocation to Lake Chapala are particular weaknesses. Monitoring and ensuring the quality of primary water sources are relatively weak functions, and controlling groundwater pollution is not covered at all, as is the case in the Gediz basin in Turkey. Advocacy groups do not have an important voice in basin management decisions.

In a backlash against restrictions imposed by the Council on agricultural water use, agricultural interests have recently coalesced to strengthen their voice and influence. WUA presidents from five basin states have met together to coordinate their actions and have formed a new Working Group for agriculture within the Basin Council. Interestingly, as in California and Turkey, limitations on surface water access have led irrigators to increase exploitation of less regulated groundwater, exacerbating a groundwater overdraft problem in all three countries.

To try to regulate this overdrafting, member states of the Council signed a coordinating agreement in 1993. To date, though, there is little to show. While the State has prohibited the sinking of new wells in the basin, groundwater development continues apace, and overdrafting accelerates.

Another issue to be faced in the future is the right of access to derivative water. With a major programme of constructing wastewater treatment plants now showing results, raw wastewater streams, which were available for reuse in rivers heretofore, are now controlled and improved in quality and can be diverted to new users who are willing to pay for the right to use them. Complicated issues of rights to derivative water and compensation for lost use rights await.

12.3.1.4 Olifants

Prior to majority rule in 1994, water resources in South Africa were managed for the benefit of the white minority under a hybrid system of riparian and appropriative rights. The national Department of Water Affairs (DWA) was the pre-eminent organization managing surface water and wastewater, playing a major role in planning, allocating, supplying and protecting surface water resources.

After 1994, water institutions, along with most other public institutions, underwent massive fundamental changes. The 1996 constitution gave 'basic rights' status to access to water to support life and for personal hygiene and a healthy environment. There followed two new water acts in 1997 and 1998, which remapped the water landscape. Water was made a public trust and the Department of Water Affairs and Forestry (DWAF) was made their custodian, with wide-reaching powers similar to those of the CNA in Mexico.

Following 1994 and in accord with the new water legislation, institutions that formerly served white farmers and communities were modified and expanded to cover all citizens. Thus, for example, the former irrigation boards are being converted into WUAs and established wherever irrigation will be practised across the country, arrangements similar to those existing in the USA, Mexico and Turkey.

In a major break with tradition, the management of water resources was assigned to a set of 19 semi-autonomous Catchment Management Agencies (CMAs) to be created in major water basins in the country. The Olifants is one of these. The CMA must plan a 5-year management strategy and will progressively assume management and monitoring functions delegated to it by DWAF. It will collect fees from water users to cover its costs. The roles envisioned for the CMA span a wide range, including planning and water allocation, wastewater regulation, and monitoring and enforcement of water quality standards. It remains to be seen whether all of these functions, particularly the enforcement ones, are appropriate to such a pluralistic institution.

The process of forming a CMA in the Olifants basin has been deliberate, with much attention paid to obtaining widespread participation of basin stakeholders in the formative process. However, with more than 3 million people in the basin, securing the vitally important participation of black rural residents has been difficult. Other interests, such as industrial users, power generators and white commercial farmers are well organized, articulate (in English and Afrikaans) and politically adept, and tend to overshadow smallholder farmers and other rural residents.

The process of creating locally accountable basin management organizations is in an early stage. It is underlain by an exemplary body of legislation, a wealth of good intentions and capable technical support, and faces an enormous challenge in bringing stakeholders together to govern the CMA in a fair and balanced way. The technical challenges of information collection and management, planning, and reallocation also lie ahead.

12.3.1.5 Dong Nai

Like Mexico and South Africa, Vietnam also has a new and comprehensive water law. Promulgated in 1999, the new law provides a forward-looking structure for the sector and mandates that water resources be managed on the basis of hydrologic boundaries rather than administrative ones. An apex-level National Water Resources Council was established in Vietnam in 2000 to anchor the new set-up. While logical and consistent with international thinking on the topic, the notion of management at the basin level across administrative boundaries runs counter to the administrative model currently in place, wherein administrative boundaries form the basis of a complex, interlocking and hierarchical administrative set-up extending downward from the national level to provinces, districts and communes. Administrators at all of these levels are currently involved in water resource management, and it makes for a very complex situation.[1] Unfortunately, too, the tendency is to see basin-level management as an overlay to this existing structure rather than as a new model that cuts across and supplants existing governmental units and authority, and this has the potential to make the set-up even more complex.

As in South Africa, the government is only just starting to implement this new model. A Planning Management Council for the Dong Nai River basin was established in 2001 under the new water law. The 12 permanent members of the council are all representatives of national level ministries and agencies, while the involved provinces are represented by non-standing members drawn from more junior administrative ranks. Moreover, provincial representation is drawn from the provincial Agricultural and Rural Development Service and does not include the sectors of public health, domestic water supply, wastewater handling and treatment, and environmental quality. It seems likely that the initial role of the Council will be planning new basin development projects, serving as an extension of the Sub-Institute for Water Resources Planning, which also acts as the Council's secretariat.

A parallel basin-wide coordinating effort has been ongoing for several years under the sponsorship of the Ministry of Science Technology and Environment. Participants are trying to establish a ten-province-plus-HCMC committee to monitor and manage water quality in the basin. This effort differs from the MARD-based Planning Management Council in that its primary members are provincial governments and in its focus on water quality issues. For the time being, these two basin-level organizations appear positioned to evolve on parallel tracks. In some ways, this is a positive development as it aids the emergence of a provincially based capacity to engage in basin management activities in addition to the nationally based one. Coordination will be required, and at some point, the two mechanisms may be merged, but this should not be done at the expense of the more democratic and representative process emerging at the regional level.

12.3.1.6 Gediz

The Gediz basin represents the negative image of basin management – obvious in its absence. The national water agency, DSI, is similar to the Mexican CNA in its broad and comprehensive powers to develop, manage and regulate the nation's water resources. It has been in existence for 50 years and is a capable and professional organization. As with the CNA in Mexico, DSI has implemented a very successful management transfer programme over the past 10 years, shifting management responsibility for 80% of the irrigated area under its charge to locally controlled Irrigation Unions.

However, despite its professional competence and its success in devolving operational control of most of its irrigation systems to water users, DSI has not yet re-interpreted its mandate to include basin-level water resource management. As a result, there is a sizeable gap in institutional coverage of basin management functions.

This is manifested in very limited planning of surface water at the basin level and virtually none for groundwater. There

---

[1]   The actors and functions matrix for the Dong Nai basin is considerably larger than those for each of the other cases studied.

is no integrated plan which considers both ground and surface water availability, nor does existing planning consider water quality, wastewater disposal, current and projected land use, anticipated future demand and return flows, or projected future quantity and quality of water resources.

Water is allocated, in practice, by a variety of agencies and users operating independently of each other. There is no national legal framework for surface water rights and only a rudimentary system of allocating access to groundwater. As a result, both are largely open-access resources. Although nominal control is stronger for groundwater than for surface water, groundwater is presently the more stressed of the two resources. It is also the more desirable of the two, in part because of the poor quality of Gediz surface water in the lower reaches, but also because of the relative ease of access provided by groundwater. The current system of registering groundwater withdrawals has been no more effective at limiting overdrafting than practices in California and Mexico.

Water quality monitoring takes place, but information is often not available in useful forms to interested parties. The information collected remains as raw data in agency files and is not generally used as a basis for policy making or decision making for basin management.

Despite serious quality degradation of surface waters in the basin, there is little enforcement of water quality regulations because a single official, the provincial governor, is empowered to authorize enforcement. Although a number of environmental NGOs exist in Turkey, these have not yet developed sufficient confidence and independence to lobby effectively for enhanced environmental protections.

In sum, the need for basin-level management in the Gediz is acute, and adequate authority and professional competence exists within DSI to begin this process. Bringing the process to completion will require a new national system of water rights to safeguard existing users and prevent over-allocation; however, there are many intermediate steps related to integrated planning

and data collection and dissemination that could be taken now. This has not been a high priority for DSI, perhaps because of a primary focus on extensive irrigation construction activities elsewhere in the country.

12.3.2 Enabling conditions

Additional attributes of well-functioning basin management systems relate to dynamics. We term these attributes enabling conditions (see Chapter 1). Enabling conditions are features of the institutional environment at the basin level that must be present, in some measure, to achieve good governance and management of the basin. These attributes are not specific to any one actor, but apply to all actors and their interactions and comprise necessary, but not sufficient normative conditions for success.

12.3.2.1 Political attributes

An important political attribute of an effective basin management system is adequate representation of interests. Equally important is the need for a rough balance of political power and influence among the various interests. When power is one-sided, issues are not aired adequately, and decisions are also one-sided.

In two of the study cases, a wide array of interests represent themselves in basin management discussions and decision making. In California, the various principal sectors – agricultural, municipalities and the environment – are well organized and funded and represent their members in a variety of different forums. Involved entities include associations of irrigation districts, associations of municipal water districts, federations of farmers, state and federal management and regulatory agencies, and environmental NGOs. Following debate, decisions are made by the state legislature, the US Congress and regulatory authorities. Disputes are adjudicated by state and federal courts. Voluntary agreements among parties are often reached after negotiation and formalized in legally enforceable contracts.

In France, in a simpler set-up, the Basin Commission or 'water parliament' serves as a discussion and decision-making forum involving the major actors.

In three other cases, those of Mexico, South Africa and Turkey, some interests are well organized and well represented, while others are not. For example, in the Gediz, local industrialists form associations, which promote their interests with DSI and the government. Farmers, though more numerous, are less well organized and funded, and do not speak with a unified voice. However, Irrigation Unions recognize their disadvantage in this regard and have begun to discuss coordination and possible federation among themselves. In Mexico also, the success of WUAs in five Lerma–Chapala basin states in forming an irrigation working group within the Basin Council and in taking a common position against further reductions in their water allocations shows their increasing sophistication and an evolution toward greater balance in representation. In both countries, the environment is poorly represented in debates over allocation and pollution because of the relatively weak status of the national environmental agencies and the relative absence of effective environmental advocacy groups based in civil society. Neither country appears to be taking a proactive stance to rectify these imbalances.

In South Africa, the cleavages are quite different. Here, the dualism resulting from the apartheid legacy has created a situation where wealthier white segments of all sectors of the water economy are well organized and represented, while the generally poorer black segments of the same sectors are not. Industrialists and commercial farmers are better endowed to participate in consultation processes and are concerned about continued access and water quality issues. The environment is well represented and supported by strong provisions in the National Water Act mandating priority allocations to meet environmental needs.

Administrative structures serving black South Africans in the former homeland areas were extremely weak and have been disbanded. These populations currently have to grapple with major ideological differences between traditional tribal leadership and the new democratically elected representation. This disparity in representation and influence is rightly recognized by policy makers as the pre-eminent challenge in developing democratic and representative basin management agencies.

They recognize that the new system must create mechanisms through which water users across the board can make themselves heard and understood, enabling gradual and systematic redress of racial and gender inequities while ensuring a secure base for economic growth. Lack of access – both through lack of rights and lack of infrastructure – are priority issues for the rural poor in the Olifants basin. At the same time, however, adequate representation is hard to achieve: large numbers of people live in remote areas, excluded through the cost of transportation and an absence of organization. The government, through DWAF, is experimenting with a variety of approaches to achieving better balance in representation.

In Vietnam, the model is quite different. Power rests overwhelmingly with the State and the Party, and there is scant scope for representational associations based in civil society. It is assumed that the various organs of government – ministries of agriculture, environment and energy, and the multitude of other involved government bodies – can adequately represent the interests of all involved stakeholders. Moreover, national level agencies tend to have much stronger representation than provincial and local level ones. As a result, basin forums tend to involve only public sector actors and are dominated by those representing State (national) agencies. The result is an administrative technocratic approach to planning and management which does not necessarily reflect the values of provincial authorities or private economic actors active in the various sectors.

In sum, in the five multi-party democracies, representation of the variety of public and private interests involved in basin planning and management is either present or evolving. The modes and mechanisms for this vary widely, as do the present balances among interests, but the involved stakeholders themselves, and, in most cases the governments, recognize the importance of this process. In the sixth case, State organs represent all of the involved interests, resulting in a more centralized, public policy-oriented and technocratic approach to planning and management.

12.3.2.2 Informational attributes

Another essential enabling condition is the presence in the public domain of accurate and up-to-date descriptive information on water in the basin, along with open and accessible decision-making processes related to plans, regulations, violations and sanctions. The first of these stipulations requires that information on basin water allocations, reservoir positions, groundwater elevations, water quality conditions, available resources and so on be a part of the public record. This disclosure condition applies to intra- and interdepartmental information relationships as well as to those with the general public. The second stipulation, transparency of public proceedings, is similarly essential to fair democratic processes. Rent-seeking behaviour requires darkness and privacy to thrive, and conducting regulatory processes in full view of the public is an effective antidote to such practices.

With regard to data openness, the contrast between California on the one hand, and Turkey and Vietnam on the other is strong. In California, law requires such openness and the many players in basin management demand it. Information is readily available in paper form and on the Internet. In the other two cases, information is often regarded as a source of power and is jealously guarded by the collecting unit. There may be a reluctance to share even within the same agency or ministry. Such sequestering of information may be rationalized in terms of state security, but legitimate justification for this is usually lacking, and the real reasons generally relate to control, power and perhaps to avoid conflict or embarrassment.

Generating improved information on basin water resources and making it available to all interested parties can be an early step in establishing basin planning and management mechanisms, and can be applied in both democratic and technocratic environments. Improved information gathering and dissemination can help to ease tensions between involved parties, even where distrust is extreme. In the Kura–Araks basin of the South Caucasus, for example, where the three riparians of Armenia, Azerbaijan and Georgia agree on little else, they have recently been willing to collaborate on data collection and sharing relating to river hydrology.

In South Africa, the problem is different. One of the lessons of the water law consultations was the importance of trusted information as a basis for consultation and negotiation. Good information is crucial to delineate areas of agreement and disagreement and to structure and inform debate, but of little use if the source is doubted. Equally, good information becomes useful in negotiation and decision making only when it is accessible by all interested and affected parties. South Africa is well equipped to use the most modern techniques for data gathering, storage and knowledge creation, but faces a major challenge in presenting information in a meaningful way to the wide range of interests in the sector. Those most in need of water for basic and productive uses are poorly equipped to access and interpret information from the national data collection systems. Good information has the power to defuse unnecessary tensions, while misinformation and lack of information foster conflict. In Mexico also, good information is available, but other actors in addition to CNA, which collects most of it, need to learn to assess, analyse and employ it. Ultimately, private stakeholders may begin their own programmes to collect and analyse key information as a check on information from official public sources.

12.3.2.3 Legal authority

Mexico, South Africa and Vietnam all have recent comprehensive water laws to serve as a basis for water resource management. The other three case study countries rely on

a pastiche of older laws to guide and structure the sector. There does not seem to be
any particular relationship between the
existence of a modern comprehensive water
law and successful basin management
practices. In California, for example, some
legislation governing the sector, such as
that governing water rights, is more than
100 years old, while laws dealing with
other aspects are quite recent. In addition,
court decisions in California may have the
effect of modifying law and creating new
principles. For example, the public trust
doctrine was added to the body of California law by a court decision in 1983. South
Africa achieved this same effect through
provisions in the new water law.

In Turkey, the 1954 law establishing the
national water resource agency consolidated
broad authority over water resources in that
agency, which has allowed it to respond to
many of the changing needs it has encountered over the past 50 years. For example, it
was able to use another existing law autho-
rizing local governments to form consortia to
deal with issues cutting across their boundaries to establish effective Irrigation Unions
in the 1990s. However, the absence of a
legally sanctioned system of water rights,
poorly designed enforcement powers for
water quality standards, and reform of the
structure of the Irrigation Unions all require
new legislative action, and the absence
of such reform legislation hamstrings the
development of basin water management
mechanisms.

In Vietnam, the principles embodied in
the new water law still await implementing
rules and regulations to translate them into
actionable practices. Effective implementation also requires establishing administrative systems for managing water permits,
collecting and consolidating data, and the
like. In a larger sense, too, to be fully implemented, the tension between principles in
the new law and the prevailing administration culture will have to be resolved.

The important lesson in this is that
developing and promulgating a new and
comprehensive water law to subsume and
replace existing laws is not a necessary condition for effective basin management. This
step may be useful when a major transformation is contemplated, as it was in South
Africa and Mexico, but it does not appear
essential. What is important, however, is
that the legislative basis for water resource
management be kept current through regular
legal amendments, updated administrative
rules and reinterpretations of current rules
and practices. Where this has not happened,
as in Turkey, emergence of new basin
management institutions will certainly be
constrained.

12.3.2.4 Resources

Resources needed to practise basin
management can be classified as human,
financial, institutional and infrastructural.
The human resources needed are those
individuals skilled in engineering, hydrology, water chemistry, economics, mass
organization, management and other
disciplines needed to plan and manage
the use of basin water resources. In all of
the cases studied, professional skills in the
basic disciplines are probably available,
though particular skills, professional
management expertise for example, may be
in short supply in particular countries. In
addition, specialized training will often be
necessary to build on basic disciplinary
expertise. This will be particularly true in
countries just beginning to practise basin-
level water management. To a degree also,
the availability of these skills to the basin
organizations will depend on the financial
resources available to hire them, and to
compete with other potential employers in
cases where particular skills are scarce.

As noted earlier in this chapter, the
wealth of a society will help determine
the financial resources available for basin
management tasks. Most funding is usually
drawn from the public treasury, since tasks
such as data gathering and analysis, issuance and recording of water and wastewater
permits, monitoring, and enforcement of
rules and standards are public sector tasks.
However, it is also important that some
resources be available from private user
groups as well to preserve their independence. In France and California, for example,

representational resources for different user sectors are drawn from the members of these organizations. In Mexico and South Africa, this is happening for some, but not all, user sectors. Within the public sector as well, it is important that resources be available through the budgets of different sectoral ministries and agencies, rather than being routed through a single ministry budget, so that potentially divergent interests such as hydropower generation and flood control or agriculture and environment can be represented with some independence. On the other hand, a potential problem is scattering of financial and human resources among too many organizations, where each lacks a critical mass to be effective, as may be the case in Vietnam at present.

Institutional resources are those needed to translate policy and law into action. They include public organizations tasked with collecting and processing data and monitoring and enforcing standards, user organizations to represent sectoral interests, and coordinative and adjudicatory bodies to knit sectors together and reach decisions. In a broader sense, institutional resources extend to cover basic social institutions such as rule of law, enforceability of contracts, right to organize, and so on. These are less amenable to change than more specific institutions are, such as a system of water rights, but are important determinants of the type of basin management set-up chosen and the expectations held for it. The stock of institutional resources varies widely among the case study countries.

Infrastructural resources are those that facilitate control of the water resource. They include plumbing such as dams, diversions, canals and reservoirs, as well as measurement and data management and processing technology. One of the most important of these characteristics is the amount of controllable storage available in a basin, both above and below ground. This characteristic interacts with basin hydrology, since snow-fed rivers such as those of the Central Valley will need less storage, other things being equal, than those fed by monsoonal rainfall such as Dong Nai. Without storage, options available to managers are much more limited. The storage estimates shown in Table 12.1 for the various basins show the wide range of controllable water supply available to managers in these basins and suggest different strategies for dealing with basin development and management.

In addition to the ability to store water, infrastructural resources also allow transferring water from one basin to another in response to politically expressed demands and relative water availability. This capacity is considerable in the Central Valley and Neste, and is also present in Olifants and Dong Nai. This capacity often, but not always, comes in later stages of basin development.

12.4 Conclusions

Conclusions from any set of case studies always depend on the cases making up the set, the information included in them and the frame of reference employed in analysing them. Recognizing these limitations, we offer the following conclusions drawn from our interpretation of this particular set of cases and tempered by our experience in other basins around the world:

- Constrained surface water availability, coupled with an absence of readily accessible groundwater to take up slack, is the most powerful driver of concerted water resource management at the basin level. Other needs which constitute potential drivers of basin level management – overdrafted groundwater, water pollution and environmental degradation – are usually less potent than constrained water availability in fostering the necessary institutions. In part, this is because excess surface water can compensate for water quality degradation by diluting waste streams and substitute for constrained groundwater, though perhaps at higher cost.
- High-quality basin management services are expensive, and wealthier societies are better able to bear these costs than poorer ones. Moreover, wealthier

societies often have a broader range of human and institutional resources available internally that are needed in basin planning and management than do poorer ones. This does not prevent the practice of basin management in any society, but may influence the form that management takes and the objectives established for it.

- Basin management, as conceived here, is better described as a political process than a technical one. This observation stems both from the concepts of 'rules in use' and 'water as a contested resource' developed in Chapter 1, and from practices, organizational structures, and interactions described in the six case studies. Going a step further, basin management is, in essence, a democratic process, rather than an autocratic or administrative one. Although technical processes of water measurement, demand estimation, modelling, water quality sampling, and so on are involved, the core processes of basin management involving making fundamental decisions on water allocation and quality on the basis of particular economic interests, social values, priorities and expectations – decision making that is inherently political. In this conception of basin management, pride of place belongs to the governance of basin management processes, in turn placing high value on access to information, transparency, accountability, balanced representation, and rule of law. Introducing basin management processes into a social context which also values these characteristics of good governance is consequently much easier than introducing them into one which does not.

- Management institutions do not have to be embodied in a unitary organizational structure to be effective. A more dispersed set of organizations can also manage a basin effectively if they are knit together with suitable processes, rules and other institutions, and if they provide for revision and updating. That said, when a country chooses to make a

radical change in its organizational set-up for managing water resources, a comprehensive reassessment and restructuring may be an appropriate way to understand and introduce the interlinked set of changes that such a major reorganization implies.

Finally, we offer for consideration a set of four broad institutional changes that seem to take place as a basin closes. The first of these is expansion in the number of public and private sector actors involved in basin planning and management. This expansion includes service providers for water uses of growing importance, such as municipalities, industries, environmental agencies; citizens groups; and other regulatory bodies. In addition to growing scarcity, this expansion is occasioned by shifts in the composition of demand for water, deteriorating water quality, growing public concern over environmental quality, and growth in public wealth and increased professionalism in resource management as a result of overall economic development.

A second change is that organizations associated with basin planning and management become more specialized and differentiated. One important shift that typically takes place is the separation of basin functions into three broad categories:

- Regulation and standard setting;
- Resource management;
- Service provision.

Regulation and standard setting are carried out in the public interest and are necessarily functions of government. Resource management tasks include allocating and protecting water resources, and must be done even-handedly and responsively to both clients and regulators. Resource management is generally done by public agencies, or by hybrid public/private organizations. Water-related services such as irrigation and drainage or municipal water supply are generally provided by a special-purpose organization operating under private-sector-like incentives to provide services to clients effectively and efficiently. Experienced observers argue convincingly, in fact, that such differentiation is a necessary precondition of a functional and efficient basin management structure as closure is approached (Millington, 2000; van Hofwegen, 2001).

The third shift is greater involvement of civil society in basin planning and management. With rising standards of living, urbanization, and in the face of progressive environmental deterioration, elements of civil society increasingly mobilize to protect the environment and reverse cumulative adverse effects of resource exploitation and industrialization on it. By organizing into associations, societies and other groups, this perspective gains entry into water resource decision-making processes.

The fourth shift is toward a broader range of disciplines playing roles in basin planning and management. A typical progression begins with engineers and hydrologists and then expands to include economists, agronomists, management specialists, ecologists, public health specialists, water chemists and others. This places a growing premium on communication and sharing of information, both among specialists and with civil society as a whole.

These changes seem to occur in a variety of contexts as water resource exploitation progresses, but they are in no way ordained by basin closure. Basin closure is just one of a set of changes that proceed simultaneously as societies grow and develop. These changes are interlinked, but not rigidly so. Corollary changes in a society include expanding exploitation of all available resources (including water); rising agricultural productivity; expanding industrial and service sectors; a growing population; urbanization, rising per capita incomes; rising educational levels; burgeoning waste discharges to land, air and water; and numerous others. Growing scarcity of water creates pressures and incentives for institutional change, but the actual manner and pace at which these institutional changes take place are particular to the prevailing social, political and cultural environment. Moreover, in the absence of leadership and direction, countries may 'sleepwalk' into potential disaster as basin closure progresses, but without appropriate institutional responses. Moreover, the pace of institutional change will accelerate as closure approaches, i.e. more rapid change will tend to occur between stages 2 and 3, utilization and reallocation, than between stages 1 and 2, development and utilization.

References

Millington, P. (2000) River basin management; its role in major water infrastructure projects. Prepared as an input to the World Commission on Dams, Cape Town.

van Hofwegen, P. (2001) Framework for assessment of institutional frameworks for integrated water resources management. In: Abernethy, C. (ed.) *Intersectoral Management of River Basins*. IWMI, Colombo, Sri Lanka.

13 Providing Irrigation Services in Water-scarce Basins: Representation and Support

Philippus Wester, Tushaar Shah and Douglas J. Merrey

13.1 Introduction

As water becomes scarcer, competition intensifies and its value rises; irrigated agriculture must change profoundly to use water more productively. Compounding this challenge is the widespread poverty in river basins in developing countries and the pressure this creates to reallocate water to the poor for productive uses. This chapter draws from the river basin studies in this book, along with supplementary materials, to outline the challenges facing irrigation service provision in closing river basins in developing countries and the implications this has for institutional support systems. It also addresses the adequate representation of the large numbers of small-scale water users characteristic of developing country river basins, along with other interests, in basin-level decision-making institutions.

In semiarid regions, irrigation is generally the largest water user in the basin. In the past 50 years, vast sums of money have been invested, by both governments and farmers, to construct new irrigation systems or rehabilitate deteriorating ones. The irrigated area nearly doubled from 140 million ha in 1960 to 275 million ha in 2000. In many countries, this occurred under the banner of integrated river basin development, aptly summarized as 'the orderly marshalling of

water resources of river basins . . . to promote human welfare' (UN, 1970, p. 1, quoted in Barrows 1998, p. 172). In the past 15 years, governments have shifted away from irrigation infrastructure development for poverty reduction to improving irrigation management, frequently through institutional reforms such as irrigation management transfer (IMT). An overriding concern in irrigation reforms has been to reduce government subsidies to the operation and maintenance of irrigation systems, while the focus on poverty reduction through the development of large-scale irrigation has waned. Only recently has a paradigm shift started to occur in water management, based on the recognition that providing access to water for drinking and growing food, eradicating poverty, and stopping groundwater overexploitation are central challenges in river basins in developing countries that require new ways of thinking.

In the developing world, institutional arrangements for basin management are only starting to take shape and are still rather weak at a time when the management of river basins is becoming more difficult and challenging (Vermillion and Merrey, 1998). In part, this is due to the widespread trend to decentralize the management of water services, such as with IMT or to privatize urban water supply, but also because local institutions – such as water users

©CAB International 2005. *Irrigation and River Basin Management* (ed. M. Svendsen)

associations (WUAs) – are young, weak and dependent on agencies for resources as well as technical support. While having mixed outcomes in terms of accountability and cost effectiveness at local levels, decentralization leads to a multiplicity of organizations, often with competing interests, and makes water management in river basins reaching closure more challenging. Although a large body of literature has developed on river basin management, relatively little attention has been given to the needs of locally managed irrigation and how to address water deprivation in closing river basins in developing countries. The closure of river basins (see Chapter 2) in semiarid regions poses significant challenges for pro-poor water policies. The following issues stand out:

- The over-exploitation of primary water sources (waters tapped from rivers, lakes and aquifers, i.e. 'blue water') leads to environmental degradation through the destruction of aquatic ecosystems, depletion of aquifers and generation of polluted wastewater flows (both industrial/urban effluents and agricultural drainage effluents). In closed river basins, the only way to reverse these trends is to consume less primary water and to make judicious use of derivative water (municipal wastewater, industrial discharges and agricultural return flows).
- Alleviating poverty through the creation of new hydraulic property (Coward, 1986) becomes very difficult as primary water sources are already fully committed, and frequently under the control of the relatively better off. Creating new entitlements for the poor must therefore be sought in renegotiating rights to primary water. Equally important is the need to increase the productivity of 'green water' (water stored in the soil profile) through rainwater harvesting and anti-erosion measures.
- The lack of possibilities to develop new water supplies and perceptions that agriculture is a 'low-value' use of water leads to increasing inter-sectoral water

transfers. These are frequently one-way transfers from agriculture to industry and domestic use, as well as intra-sectoral transfers in agriculture to higher-value crops, usually grown by commercial farmers.

- Without clear water rights and effective enforcement, it is relatively easy for poor people, such as smallholder irrigation farmers, to lose access to water for production due to these transfers. Consequently, poor farmers increasingly have to turn to derivative water as their only source of water.
- Before IMT, most irrigation schemes were under government management and to a degree, protection. After IMT, this protection may disappear, and the local irrigation management entity may find itself under-represented at higher levels (Svendsen *et al.*, 2000).

To address these issues, concerted change at different levels, building on the strengths of government, civil society groups, popular movements and communities, is necessary. Such change needs to focus on the creation of interlocking institutional arrangements at the local, meso and macro level to manage water in a socially just and equitable manner that meets the needs of the poor and ensures the sustainability of locally managed irrigation. Of special concern is the need to represent and protect the interests of water users that are at risk, such as the large number of small-scale water users that fall outside the ambit of formal water management. Furthermore, the ability of groundwater irrigators to make mincemeat of any effort towards orderly resource management at the basin level needs to be taken into account. Previous chapters of this book have analysed the institutional challenges that arise when the utilization of water nears or exceeds the annual renewable water in a river basin and how various countries are dealing with these challenges. In this final chapter, we endeavour to draw lessons from these experiences as they relate to irrigation service provision and poverty alleviation. One fundamental premise of the chapter should be stated explicitly: in poor developing countries, it is desirable to find ways to preserve and improve irrigated agriculture, as this is a major source of livelihoods and food security for which there are few substitutes.

13.2 Challenges Facing Locally Managed Irrigation in Closing River Basins

Between the 1950s and 1980s, vast investments were made in poor countries to increase the area served by large-scale irrigation systems dependent on surface water. A central purpose of this effort was to reduce poverty and to attain national food self-sufficiency. However, starting in the 1980s, serious concerns were raised that public investments in large surface irrigation were not sufficiently benefiting poor farmers and that the government agencies charged with irrigation management were performing poorly, especially in recovering costs from farmers. To reduce the burden on the public purse, and under the influence of the structural adjustment programmes of the 1980s, the role of government in irrigation management in many developing countries started to change, with an emphasis placed on transferring management responsibilities and financial obligations to farmers. Much of this reform is captured in the phrase 'irrigation management transfer', though the actual reforms are broader than 'simply' transferring water management responsibility from government to water users. The widespread trend to devolve irrigation management to farmer organizations, coupled with the rapid increase in groundwater irrigation and water harvesting and local storage means that currently a large majority of the world's irrigation is locally managed. A distinguishing characteristic of large portions of locally managed irrigation is that it has thin or no contact with formal resource governance structures. Examples include the 20 million people pumping groundwater in South Asia and the communities that depend on South India's 300,000 tanks or China's 7 million ponds.

The chapters in this book have shown that there are clear stages to river basin development related to the changing pattern of demand for water over time. A fundamental point for understanding the challenges facing locally managed irrigation in water-scarce basins is that irrigation reforms such as IMT were not enacted to deal with river basin closure but rather to reduce government expenditure. Thus, management transfer in many countries is occurring at a time when water resources in river basins are becoming increasingly scarce and the subject of inter-sectoral competition. This places an extra burden on farmers, who, while having had little time to develop their associations, immediately need to start focusing on the river basin level to secure and retain an adequate share of water. As many governments have been transferring irrigation management to user organizations rather hastily, the world's irrigation is increasingly likely to be managed by ill-formed and ill-prepared user organizations. Where systematic efforts are not made to support newly established WUAs and secure their access to water, the sustainability of smallholder irrigation itself may come into question (cf. Shah *et al.*, 2002). At the same time, burgeoning informal irrigation economies, unrestrained by national policies and government regulatory structures, defy initiatives for more orderly basin-level water allocation and management, and are instead driven by their own rules and internal logic.

The challenges facing locally managed irrigation in water-scarce basins are thus twofold: internal and external. This chapter mainly deals with the external challenges, as previous studies have adequately identified the internal support needs of locally managed irrigation (Yoder, 1994; IIMI, 1997, 1998; Frederiksen and Vissia, 1998; Huppert and Urban, 1998; Svendsen *et al.*, 2000; Huppert *et al.*, 2001). Local irrigation management entities need to focus internally on improving irrigation water management, while at the same time negotiating externally with policy makers, river basin authorities and other water users to protect their water allocation at the basin level.

They must become more outward-looking to gain or maintain access to water supplies that other sectors may try to capture. As the entities managing local irrigation are spatially dispersed and mainly focused on their own irrigation system, they need to confederate to lobby and compete at the basin level. Compounding this challenge is the declining share of the agricultural sector in many countries, and the perception that other uses of water are higher value, while the persistent dependence of a large share of the population, especially of the poor, on farming for livelihoods continues. There is growing pressure on local irrigation management entities and farmers to increase water productivity and to make the case that they are not wasting water.

The above challenges apply to both surface water and groundwater management organizations, and locally managed irrigation in upper catchments as well as newly formed water users' organizations in large-scale irrigation systems under IMT programmes. Although this is necessarily a broad-brush approach, and the specific challenges to the myriad irrigation organizations that exist depends on many more factors than river basin closure, several common threads emerge from the analysis of the basins studied in this book.

13.3 Irrigation Service Provision in Water-scarce Basins

The challenges facing locally managed irrigation in water-scarce basins outlined above directly affect the institutional requirements for irrigation service provision. A service provision perspective on irrigation management entails focusing on the different service roles or functions performed by the multiple actors involved in water management in a river basin and the mechanisms that govern the exchange of services between them. Huppert *et al.* (2001, p. 41) distinguish between primary services (the provision of irrigation infrastructure and water delivery), secondary

services (maintenance and operation of infrastructure) and supporting services. Of importance in irrigation service provision is the large number of actors involved, as the provision of primary services (such as water delivery to a farmer) is the result of a network of (supporting) service providers and receivers. Thus, when analysing service networks, it is necessary to look at the service relationships between actors by identifying the laws, procedures, contracts and/or common practices that are the basis for the relationship, i.e. the governance mechanisms between service providers and receivers. Analysing service arrangements, whether highly formalized in contracts or based on customary understandings, entails studying the following:

- Which services are being provided, by whom and to whom;
- What is being exchanged in return for each service;
- Which governance mechanisms structure the service delivery and whether this results in the provision of services in a way that suits those concerned.

Our concern here is not with the service arrangements between farmers in the case of self-provision of services, which is the most common form in much of the developing world, or with the arrangements between farmers, local irrigation management entities and others governing the provision of primary and secondary irrigation services. While important for a clear understanding of the internal workings of locally managed irrigation and identifying areas for improvement (see Huppert *et al.*, 2001, for different strategies to strengthen locally managed irrigation), we focus here on the external threats to the sustainability of locally managed irrigation. Several essential support needs for locally managed irrigation in water-scarce basins stand out, namely water rights systems and water allocation mechanisms, compensation mechanisms for water transfers, and increasing water productivity. These three support needs are briefly outlined below.

13.3.1 Water rights systems and water allocation mechanisms

The basins studied in this book demonstrate the increasing pressure being placed on irrigated agriculture to relinquish water, primarily for environmental and urban/industrial uses. It follows that in water-scarce basins with increasing competition for water, the need for effective mechanisms for allocating water becomes critical. In the absence of such mechanisms higher-value uses will tend to out-compete lower-value uses (with value being defined politically as well as economically), depriving them of water and leading to unregulated transfers of water out of agriculture. A crucial support need of locally managed irrigation is basin-level water allocation mechanisms for both primary and derivative water, which are based on defined water rights that provide security of tenure. The call for clear, secure and transferable water rights has often been made, but how to create property rights to water that are just, equitable and feasible (both technically and politically) and how to make them stick is neither clear nor straightforward. Nonetheless, several principles can be defined.

Water rights form part of property regimes, which define ownership and consist of principles and rules to resolve disputes over property. They may be defined as 'authorized demands to use (part of) a flow of water, including certain privileges, restrictions, obligations and sanctions' (Beccar *et al.*, 2002, p. 3). Ideally, water rights delimit the amount of water a right-holder is entitled to, defined either volumetrically or in terms of shares of available supply, as well as the duties of right-holders relative to one another and to society at large, such as quality and quantity of return flows. Water rights form the foundation of water allocation mechanisms. An ideal water allocation system is characterized by flexibility in the allocation of water supplies, security of tenure for established water users, predictability of the outcome of the allocation process, equitability and fairness. Mechanisms to allocate water may take

a variety of institutional forms, and include marginal cost pricing, public (administrative) allocation, water markets and user-based allocation (Dinar *et al.*, 1997). The prevailing system of water rights significantly influences the specific allocation mechanisms available and the effectiveness of their application. However, without infrastructure in place to withdraw and distribute water, water rights remain an empty shell. Conversely the creation of hydraulic property can lead to the *de facto* creation of water rights (cf. Coward, 1986; Chapter 11).

Due to the type of infrastructure involved and the feasibility of transparent and reliable measurements, surface water may be subject to more effective allocation than groundwater. Almost everywhere in the developing world, groundwater is treated as an apertinent right to privately owned land, and where groundwater rights significantly different from these have been tried – as in Mexico – they have defied enforcement. For surface water, three broad water rights systems have developed historically in different parts of the world based on either the riparian or the appropriation doctrines (Simpson and Ringskog, 1997). All three are 'administered' systems in the sense that an authority (government or community) plays an important role in defining, allocating and enforcing rights. Other property regimes for water are also conceivable, such as a market-based system in which water rights are awarded to the highest bidder, but are much less common in practice.

Under the riparian doctrine, a user has the right to extract water from a river system for use on land adjacent to the river as long as the water is returned to the river undiminished in quantity or quality and in a manner that does not impair downstream use. From this stringent definition, it is clear that in practice the pure riparian doctrine does not exist, and that it is also not very suited to conditions of water scarcity. Nonetheless, this doctrine is used in various forms throughout the world, especially in countries with humid climates, and generally operates without any form of permit or regulatory administration.

Under the appropriation doctrine, water is regarded as a good belonging to all and held in trust by the State or a communal authority. Water use concessions or licences are issued to users by the state or the communal authority, which ensures their right to divert or store and use a certain quantity of water. Many variations on the appropriation doctrine exist, the most common two being prior appropriation and proportional appropriation. The most significant difference between these two is how they treat water shortages. Under prior appropriation, the first rights issued on a river have priority or seniority implying that rights issued later are the first to be curtailed in times of shortage (last in, first out). Under the proportional appropriation system, concessions (usually time bound) are issued to either individual users or organizations for the use of a maximum quantity of water, although the actual quantity of water that may be used in any year is adjusted to reflect water availability within the river basin. Thus, all concession holders share in any shortages or surpluses of water proportionally. For this to work well, an administrative water allocation mechanism needs to be in place that can determine annual water availability in a timely and transparent manner. Surface water rights in the basins studied in this book are based on some variation of the proportional appropriation doctrine, with the exception of California, which has a mix of riparian rights and prior appropriation.

The closure of river basins calls for a reassessment of the usefulness of the appropriation doctrine, be it prior or proportional. The prior appropriation system, while providing security of tenure, is problematic in water-scarce basins as it does not provide for sharing shortages among right-holders, nor does it allow for new entrants (Huffaker *et al.*, 2000). Furthermore, due to the seniority principle, the actual transfer of water rights is problematic, as all appropriators more senior than the buyer will need to approve the sale, while selling water may also be interpreted as proof of non-beneficial use, constituting grounds for revocation

of the water right (cf. Rosegrant and Binswanger, 1994; Bolding *et al.*, 1999; Haddad, 2000). The flexibility of the proportional appropriation system would appear to permit the application of various allocation mechanisms, including market-based ones. However, regulating large numbers of small users is exceptionally difficult and places strong demands on enforcement.

In the basins studied in this book – and others in the developing world – existing water rights systems and allocation mechanisms are not well tailored to deal with basin closure. This is clearly the case in Turkey and Mexico, whose water rights systems are generally unable to prevent or to resolve conflicts between new and existing claims, or prevent the over-allocation of water. This is even more so in stressed basins in poorer countries in South Asia and Africa. In addition, most water rights systems primarily deal with surface water, with much less attention given to groundwater or derivative water. Thus, in the case of California, groundwater is only lightly regulated and the permissive specification of rights to groundwater has led to increasing problems with aquifer overdraft. In the Gediz basin, shallow groundwater is an open-access resource, meaning that anyone with the infrastructure to tap that water can do so. In the Lerma–Chapala basin, groundwater is in effect also an open-access resource, although formally groundwater users are required to obtain a pumping permit from federal authorities indicating a maximum extraction rate. This is by far the best that any middle- or low-income country with a substantial irrigated agriculture sector has done to bring a modicum of order in private appropriation of groundwater for irrigation. The reform has helped register all groundwater users; but restricting their withdrawals to their permitted quotas has proven to be nearly impossible.

In closing river basins with significant groundwater extractions, surface irrigation systems, both large and small, play an increasingly important – often the sole – role as cheap and effective groundwater recharge systems. In effect, groundwater users pump water previously paid for by farmers irrigating with surface water, without compensating these surface irrigators. As long as this is conjunctive use by the same farmer this is not a problem, but in areas where groundwater levels have fallen sharply and only the better off can afford to continue pumping groundwater, this does become an issue for poorer farmers only using surface water. Both in Gediz and Lerma–Chapala, there are strong hydrologic linkages between surface water flows, surface irrigation and groundwater recharge. For Lerma–Chapala, Scott and Garcés-Restrepo (2001) found that approximately 50% of canal water applied to crops recharges the underlying aquifers and subsequently becomes available for pumping. Because of this recycling, half of each unit of canal water provides subsequent additional benefit as groundwater. By way of example, assuming marginal values of Mex\$1.80/m^3 for canal water and Mex\$2.40/m^3 for groundwater, with 50% recharge the aggregate value of surface water is Mex\$3.00/m^3. However, groundwater users do not pay canal water users for the recharge function they perform, while canal water users do have to pay for the full volume of water they receive, including the 50% that goes to deep percolation.

The situation surrounding derivative waters (agricultural return flows and municipal/industrial wastewater) is even less regulated. In all the basins studied, quality standards exist for return flows and wastewater, but access to this water for productive purposes would appear to be a free-for-all, with the exception of California. In the delta of the Gediz basin and in the Lerma–Chapala basin, derivative water is becoming a critical source of water for farmers. Although no allocation mechanisms exist for derivative water, in Mexico, farmers do need to go through a fair amount of trouble to establish a tenuous claim on wastewater by reaching local agreements with municipalities or WUAs (Buechler and Scott, 2000). These claims are being threatened by the construction of wastewater treatment plants and the *de facto* reallocation of treated water to other uses, such as golf courses. In closing river basins, where the better off have already captured primary water sources and renegotiating water rights is extremely complicated, derivative water rapidly becomes the poor farmer's last resort and should be recognized as such.

Without secure water rights and clearly defined water allocation mechanisms, individual farmers and locally managed irrigation systems in water-scarce basins face an uncertain future. The hydrological interactions between surface water, groundwater and derivative water make it necessary to arrive at a coherent and feasible system of water rights and water allocation mechanisms in water-scarce basins. Several issues that such a coherent system of water rights would need to deal with stand out, namely surface–groundwater interactions, return flows, water quantity and quality interactions, provisions for basic human needs and environmental flows, and lastly provisions for new entrants. The new water rights system and water allocation mechanisms in South Africa appear to meet all these needs, and could serve as an example for the other countries studied in this book if they can be implemented successfully. The establishment, modification or clarification of water rights systems requires action at the national level (legislative and executive branches of government), but ideally should consist of a process in which all interests have adequate representation. The support services required by local irrigation management entities in this regard are legal advice and representation as well as lobbying capabilities at the basin and national level.

13.3.2 Water transfers and compensation mechanisms

A second critical support need of locally managed irrigation in water-scarce basins is compensation mechanisms for water transferred out of agriculture. This is closely tied to water rights and water allocation mechanisms, but is sufficiently important to warrant separate consideration. In recent years, there has been an increase in interest in the feasibility of establishing water markets (Rosegrant and Binswanger, 1994;

Lee and Jouravlev, 1998; Haddad, 2000). In principle, water markets enable allocation of water to high-value uses and fair compensation of those who sell their water. Especially in the case of inter-sectoral water markets, e.g. between the urban and agricultural sectors, several advantages are apparent, including more efficient use of water in agriculture and compensation to farmers for the transfer of water out of agriculture. Rosegrant and Binswanger (1994) outline the following benefits of well-designed water markets:

- Empowerment of water users by requiring their consent to water reallocation and compensation;
- Incentive for users to invest in water-saving technology, if well-defined rights are established;
- Incentive for efficient use and income through sale by considering full opportunity costs of water;
- Greater flexibility in responding to changes in the production environment (crop prices, demand patterns and comparative advantage).

While the economic virtues of efficiency and productivity are invoked as beneficial outcomes of water markets, the very real obstacles to establishing inter-sectoral water markets are often overlooked (Haddad, 2000; Young, 1986). Chief among these are issues of access to water and the lack of infrastructure to enable physical exchange and effective measurement of water (Dinar *et al.*, 1997; Perry *et al.*, 1997; Simpson and Ringskog, 1997). Market transactions assume that effective means exist for the exchange of commodities. Groundwater may be transferred from existing low-value uses to competing high-value uses only within an aquifer, not within a larger river basin unless considerable investment is made in infrastructure for water conveyance. The same is true for surface water, although the presence of rivers makes transfers from upstream to downstream uses feasible. Transferring water in the opposite direction requires substantial investments in infrastructure and control mechanisms.

Much of the claims for the 'success' of surface water markets, especially outside of the USA, rest more on political and ideological beliefs than on rigorous empirical studies (Kloezen, 1999). In the case of Chile, Bauer (1997) convincingly demonstrates that water markets are much thinner on the ground and hence less effective in water allocation than often assumed in policy papers. For groundwater, the situation is different, with active markets reported on in South Asia (cf. Shah, 1993; van Koppen, 1998). While groundwater markets have also had positive impacts on the poor (Shah, 1993), there is widespread concern that surface water markets will result in the concentration of water rights in the hands of the few, at the expense of small farmers (Bauer, 1997; Zwarteveen, 1997). The challenge this poses is to design regulated water markets that are pro-poor, recognize the social and environmental values of water, and facilitate the resolution of inter-sectoral conflicts through compensation.

If appropriately designed, water markets may provide a good mechanism to compensate farmers for water transferred out of agriculture. Pre-conditions for the effective operation of water markets include defined water rights, demand in excess of supply, legal frameworks that indicate how trades should take place, physical infrastructure for conveyance and measurement of water, and provisions for the protection of third-party interests (Perry *et al.*, 1997; Simpson and Ringskog, 1997; Lee and Jouravlev, 1998). In combination with the proportional appropriation of water rights and administrative allocation, socially just water markets are conceivable. Such a system could consist of three tiers, namely a reserve for basic human needs and the environment, a reserve for productive water for the poor and a third tier consisting of water for productive use and additional water for urban use and the environment. Only the third tier would enter the water market, after proportional administrative allocation on an annual basis has determined the water available for productive use. Special provisions would need to be made to enable the poor to lease their productive water if they so decide, while protecting their ownership of the water right. This system demands a strong regulatory and administrative framework, but is theoretically conceivable in Turkey, Mexico and South Africa, although water laws in all three countries are notably vague on the possibility of such market-based transfers.

Water rights transfers in the third tier can take a variety of forms, depending on factors such as the structure of the market, legal and third-party considerations, the characteristics of the water rights, transaction costs and above all, the needs of the parties to the transaction. Three types of transactions can be identified, namely sales, lease contracts and option contracts (Lee and Jouravlev, 1998). Sales consist of the permanent transfer of title, including all benefits, costs, risks and obligations associated with the water right. Sales are most typical of inter-sectoral transfers, with irrigated agriculture being the dominant water seller and urban users the principal buyers. The leasing of water rights involves the sale of water, but not of the water right. Leases tend to be short term, consisting of the temporary exchange of a quantity of water for monetary or other remuneration, and are frequently used for water transfers within irrigation systems.

Option contracts are long-term agreements to lease, but not to sell, a water right when a given contingency arises, e.g. a drought. Under option contracts, the receiving party holds an option to buy water at a specified price under specified conditions from the seller, who guarantees future delivery if the conditions apply. In exchange for this guarantee, the holder of the option also pays a premium to the granting party, usually at the onset of the option contract. Option contracts are most commonly used to transfer water from irrigated agriculture to non-agricultural uses such as the environment during periods of low stream flow (Landry, 1998). This means that water users can continue to use their water during years of normal water availability, and hence

option contracts are a more attractive alternative than the outright sale of water rights or long-term leases. Although option contracts are quite complex and require clearly established water rights, initial experiences with them in Chile (Thobani, 1997) and Colorado (Michelsen and Young, 1993) have been quite positive. In our assessment, an option contract arrangement may be suitable to compensate farmers for agriculture-to-urban and agriculture-to-environment water transfers. Crucial to the design of appropriate option contracts are:

- Definition of the contingency conditions under which water will be transferred;
- Duration of the contract and/or conditions for its renegotiation prior to expiration;
- Volumes to be ceded (by whom in the case of multiple parties or states);
- Specification of compensation (both monetary amount and process) for lost income resulting from the water transfer;
- Mechanisms to redress grievances on the part of ceding, receiving or third parties.

13.3.3 Increasing water productivity

The third critical support need for locally managed irrigation is assistance in increasing the productivity of water. While the view that irrigation uses water inefficiently may often be erroneous, to be credible partners in basin management local irrigation management organizations and farmers will need to show that they are using water productively. This can be measured in terms of the biomass or income produced per cubic metre of evapotranspiration.[1] Both elements of water productivity are important but place different demands on the support needs for locally managed irrigation. Higher biomass production can be

---

[1]   Not per cubic metre withdrawn, as irrigation water that is not used in the production of biomass will either recharge the aquifer or become available for re-use as return flows (see Kijne et al., 2003).

achieved through improvements in water delivery practices, crop varietal improvement and improved cultural practices (such as mulching, crop planting dates, etc). The breeding of new crop varieties, the introduction of improved cultural practices and similar actions require agricultural support services, whereas improving water delivery is the primary responsibility of the locally managed irrigation entity. However, external support remains important to improve irrigation infrastructure and management to enhance reliable water delivery and to realize real water savings.

Improving the productivity of water in economic terms is crucial for sustainable locally managed irrigation. Perhaps one of the biggest threats to locally managed irrigation is the low profitability of agriculture and the high costs of agricultural production. Solving problems related to the levels and collection of irrigation service fees as well as the quality of irrigation service delivery is an internal responsibility of locally managed irrigation entities, but the willingness and ability of farmers to pay fees is clearly related to crop market prices and economic returns to agricultural production. Where appropriate, shifting to less-water-consuming crops that provide a higher economic return needs to be supported. The efforts to do so in the Lerma–Chapala Basin, with farmers and government agencies working together to shift from low-value grain crops to crops that use less water and sell at a higher price, show how complicated it is to achieve this for substantial areas. Although increasing the profitability of agriculture touches on much wider issues than support needs for locally managed irrigation, such as trade regimes and agricultural subsidies in affluent countries, it lies at the heart of sustainable locally managed irrigation.

13.4 Poverty and Representation in River Basin Management

The case studies in this book show that ensuring the sustainability of locally managed irrigation and meeting the water needs of the poor in closing river basins is a serious challenge to current institutional arrangements. Although a service provision perspective on water management highlights where changes are necessary in the wider institutional environment to more effectively support locally managed irrigation, it is relatively silent on two critical issues in closing river basins in low- and middle-income countries, namely poverty reduction and stakeholder representation. Cross-cutting these two issues is the question of how access to water and water management decision making is gendered. A sustainable livelihoods perspective, with its emphasis on rights and entitlements and the institutional arrangements through which these are provided and reproduced, holds more promise for defining pro-poor water policies in water-scarce basins (cf. van Koppen, 2000).

13.4.1 Productive water and pro-poor water policy

Although river basin closure poses significant challenges to the sustainability of locally managed irrigation, generally this affects people who already have access to water for productive purposes. A characteristic of many stressed river basins in developing countries is the large number of poor people who do not have access to water for productive purposes. The degree of water deprivation in many river basins around the world is well documented, with more than 1 billion people lacking access to water of sufficient quality and quantity to meet minimum standards of living, let alone for productive purposes. The processes that come into play as river basins mature can have serious consequences for perpetuating poverty.

In analogy to low prices for basic grains, which hurt poor farmers but benefit the poor urban population, the reallocation of water from agriculture to urban supplies may benefit the urban poor. Whether this is the case depends on how the institutional and financial resources for urban water infrastructure development are targeted at the poor. As many maturing river basins exhibit a rapid pace of urbanization caused by rural poor moving to the cities, on balance the transfer of water out of agriculture could benefit more poor people than are hurt. Likewise, as agriculture matures and consolidates, taking water out of agriculture may be hurting the better off rather than the poorest. If farmers invest to improve the productivity – and profitability – of water there may not be any major damage to their livelihoods. Understanding how these scenarios work out requires empirical research.

However, other processes act in the opposite direction. These relate to the development of the physical means to abstract and convey water and the distribution of land and water rights. If water infrastructure development and the allocation of land and water rights is not specifically targeted at poor women and men, the danger of resource capture by the better off is ever present. While freshwater supplies are clearly limited, for most people water scarcity is caused by political, technological and economic barriers that limit their access to water and by competition between water use(r)s (Falkenmark and Lundqvist, 1998). Water scarcity is not only a naturally occurring phenomena, but has also been created through the development of water resources in the past, the selective entitlement of water rights and incidental and structural resource capture by the better off (cf. Homer-Dixon, 1999; GWP, 2000). These processes make it very difficult to increase and protect the water security of the poor, especially in closing river basins where the consumption of water by one literally deprives another of that water.

While the processes that come into play as river basins close have multiple and uncertain consequences for perpetuating poverty, it is increasingly recognized that access to water for productive uses is very important for the poor to build sustainable livelihoods (van Koppen, 2000; Schreiner and van Koppen, 2001). We maintain that pro-poor water allocation policies

can transform irrigation into a powerful instrument for creating sustainable livelihoods (cf. Hussain and Hanjra, 2004; Shah, 2000). The challenge this poses in water-scarce basins is balancing the allocation of productive water for poverty reduction with allocations designed to meet the needs of proven productive capacity (i.e. industry, commercial agriculture, mining) and the environment (cf. Perret, 2002). This may make it necessary to redistribute water rights in favour of the poor, but also calls for a judicious use of water and innovations in land and water management technologies.

To craft pro-poor water policies, an understanding of the processes that create poverty is needed. While individuals experience poverty and can work their way out of poverty, there is also truth in the statement that societies produce poverty through processes of exclusion. Culturally embedded notions of entitlements, ownership, access, control and participation underlie the concept of exclusion (Bhalla and Lapeyre, 1997, p. 417). The deprivation commonly associated with exclusion is not only related to a lack of economic resources but also a lack of recognition and entitlements. As pointed out by Sen, 'economic resources enable access not only to economic goods and services but also to political goods like freedom and the ability to influence policies' (1975; cited in Bhalla and Lapeyre, 1997, p. 418).

In this sense, access to water can be viewed as a potential vehicle to achieve economic and political rights. These are prerequisites for full citizenship, which in turn open opportunities for political participation. This interpretation brings out the state's role in exclusion. Through their structures, procedures and legal frameworks, governments can exclude some groups from fully attaining their economic rights, while including others. A case in point is the systematic exclusion of women in government irrigation programmes throughout the world, convincingly documented in numerous case studies (for a discussion of the literature, see Zwarteveen, 1994, and van Koppen, 1998).

A defining feature of poverty is that the poor have very little influence on the ways in which governments and economies allocate rights and resources in society. In closing river basins, pro-poor water policies that focus on redressing imbalances through the reallocation of water rights will challenge the existing distribution of rights and resources as there is no extra water to go around. This creates the political dilemma of confronting vested interests in society, and requires that government has both the organizational and political ability to overcome resistance to redistribution. An important first step is the formulation of water legislation that sets out procedures for the creation of a reserve of productive water for the poor and how new institutions such as river basin management entities can work in a redistributive manner (cf. van Koppen, 2000).

Of the five basins reviewed in this book, South Africa is the only country formally placing emphasis on redressing imbalances and achieving equity in water management. In the Government of South Africa's White Paper on national water policy (DWAF, 1997), three components of equity are defined: equity in access to water services (drinking water and sanitation), equity in access to water resources (water for productive purposes) and equity in the access to benefits from the use of water resources (allocative efficiency). However, even with the commitment of the government, reallocating water to new entrants is proving very difficult (van Koppen *et al.*, 2003). A way forward could be to impose water savings on commercial agriculture and other productive uses, and then allocating these savings to a water reserve for the poor. Under the system of compulsory water licensing included in the 1998 National Water Act, this is possible, but it will require determined resolve from the government to carry it through (cf. Perret, 2002). Compulsory water licensing is currently being pilot-tested in one river basin, making it premature to draw conclusions. New infrastructure will also need to be developed, to ensure that the water rights of the poor do not remain paper rights.

13.4.2 Representation in river basin management

An important challenge in river basin management is ensuring that all stakeholders have a voice in basin governance. Although frequently advocated as a key to achieving effective water management, stakeholder participation in river basin management is not straightforward, and actually including the poor and achieving substantive stakeholder representation has proven elusive in practice (Wester *et al.*, 2003). The question of how greater equity in water management and representation in river basin management can be achieved in highly stratified societies with significant gender and social inequalities remains. As poverty is an outcome of how societies are structured, it is evident that marginal groups are excluded from decision making. A danger of the emphasis on participation in river basin management is that attention is drawn away from the very real social and economic differences between people and the need for the redistribution of resources, entitlements and opportunities. In the long term, marginal groups will only gain a voice in river basin management when they are no longer marginal. This entails fundamental changes in the way societies are structured, such that they no longer produce poverty but wealth that is fairly distributed. As it is unlikely that this will happen in the near future, in the short run mechanisms need to be devised that strengthen the representation of marginal groups in river basin management.

It is clear that the size of the population in most river basins is such that it precludes the direct participation of all stakeholders in basin-level decision making. Thus, as decision making moves to the river basin level, serious thought needs to be given to how hard-won democratic rights in conventional social and political domains are assured in the river basin domain (cf. Barham, 2001). As Green and Warner (2000) point out, integrated water management and participation pull in opposite directions. While the complexity of integrated management of sizeable river basins invites centralization and technocracy, participation suggests subsidiarity and small-scale operations, engaging people to think creatively about issues with which their lives are intimately linked. Thus, in any basin of some size, river basin management would entail a layered system of representation. The question then becomes who gets to represent groups of stakeholders in river basin management. This question strikes at the heart of basin governance, and revolves around which type of democracy is implied, liberal or social.

Liberal democratic theory is premised on a notion of abstract individualism and assumes that all people are equal in the public sphere (Held, 1995; Luckham *et al.*, 2000). In water reforms informed by liberal democracy, it is assumed that it is possible for water management stakeholders to bracket status differentials and power inequalities and to deliberate 'as if' they were equals in water management forums such as WUAs or river basin councils. Social democracy, on the other hand, departs from social inequalities and attempts to increase citizen involvement in the affairs of government and expand the concept of citizenship to cover economic and social rights as well as political rights. Thus, it aims at a redistribution of power and resources to enable citizens to participate in the decisions that affect their lives (Luckham *et al.*, 2000). In water reforms informed by social democracy, water is seen as a basic human right and a politically contested resource (Gleick, 1998; Mehta, 2000).

On the face of it, stakeholder platforms, be they river basin councils, catchment management agencies or watershed councils, democratize river basin management by giving voice to a multiplicity of interested actors. However, much depends on the existing institutional arrangements from which stakeholder platforms for river basin management emerge, as many roles, rights and certainly the technologies and physical infrastructure for controlling water are already in place. In river basins, water management stakeholders may have different levels and kinds of education, speak different languages, differ in access to politics

and hold different beliefs about how nature and society function (cf. Edmunds and Wollenberg, 2001). If this is not taken into account when creating new rules, roles and rights, the institutional outcome can easily privilege those who are literate and have access to the legal system. If done unreflectively and without an emphasis on the redistribution of power and resources, new institutional arrangements for river basin management will institutionalize inequality and power differentials instead of giving voice to marginal groups (Wester and Warner, 2002). Without firm land and water rights and livelihood security, there is very little incentive for the poor to participate in river basin management.

Having argued for social democracy in river basin management, the issue of stakeholder representation remains. The relationship of the representatives participating in river basin management to their constituents is problematic, especially when third parties are involved. It is a nostrum of development work that third-party facilitators are needed to help identify, mobilize, organize and inform stakeholder groups. However, as pointed out by Edmunds and Wollenberg

> the relationship of a representative to his/ her constituency is perhaps most politically charged when representatives of a group are designated by outsiders or are accountable to them, as is often the case in multi-stakeholder negotiations. From the start, outside conveners and facilitators influence representation by the selection of stakeholder groups, the people to represent each group and how the expression of interests is facilitated in the meeting. (2001, p. 240)

This points to the need for a broad and inclusive process of invitation, consultation and consolidation of interested stakeholders, assumed to represent all interested parties. Whereas stakeholder processes and representation in river basin management are important, they need to be twinned with a focus on securing water entitlements for the poor. This points to an important role for government, both in drawing up and enforcing water laws that explicitly safeguard customary water rights and contain provisions for reallocating water rights to the poor.

13.5 Conclusion

Reflecting on the challenges facing small-holder irrigation in water-scarce basins, it becomes clear that where poverty is widespread, river basin management needs to have a strong developmental dimension. However, where the extra water for productive purposes that is needed for poverty alleviation will come from is less clear. If a country is rapidly industrializing and diversifying its economy, perhaps this is not an issue. However, most poor countries are not creating alternative employment at a sufficient rate to provide the rural poor with attractive alternatives to agriculture. In such cases trying to intensify smallholder agriculture, which often requires irrigation, is the only feasible strategy in the short to medium term. Thus, central concerns become the mechanisms in place to allocate water rights, the regulatory entity responsible for water allocation, and how conflicts about water rights and water distribution are mediated. Finding the right mix between the state, the market and the empowerment of the marginal points to the need to move beyond token consultation towards partnerships, negotiations and conflict resolution. At the minimum, strategies for river basin management should detail mechanisms for redressing imbalances in access to water for productive purposes and establishing clear and secure water rights for the poor.

While much can be learned from institutional arrangements for river basin management in affluent countries, it is crucial to understand that these do not operate in the conditions of low-income countries: dominance of smallholder agriculture, weak institutions, insufficient financial and human resources, marked social inequity and extreme poverty. While water development and management can only partly address these issues, they must explicitly form the points of departure in the reform of institutional arrangements for river basin management in developing countries. This chapter does not hold the answers to how this should be done, but it does offer elements of a strategy that could be followed to address water deprivation in closing river basins in developing countries. Such a strategy consists of a fine balancing act between allocating water for poverty reduction and allocations designed to meet the needs of proven productive capacity.

References

Barham, E. (2001) Ecological boundaries as community boundaries: the politics of watersheds. *Society and Natural Resources* 14, 181–191.

Barrows, C.J. (1998) River basin development planning and management: a critical review. *World Development* 26, 171–186.

Bauer, J.C. (1997) Bringing water markets down to earth: the political economy of water rights in Chile, 1967–95. *World Development* 25, 639–659.

Beccar, L., Boelens, R. and Hoogendam, P. (2002) Water rights and collective action in community irrigation. In: Boelens, R. and Hoogendam, P. (eds) *Water Rights and Empowerment*. Koninklijke van Gorcum, Assen, pp. 1–19.

Bhalla, A. and Lapeyre, F. (1997) Social exclusion: towards an analytical and operational framework. *Development and Change* 28, 413–433.

Bolding, A., Manzungu, E. and van der Zaagm, P. (1999) A realistic approach to water reform in Zimbabwe. In: Manzungu, E., Senzanje, A. and van der Zaag, P. (eds) *Water for Agriculture in Zimbabwe: Policy and Management Options for the Smallholder Sector*. University of Zimbabwe Publications, Harare, pp. 225–253.

Buechler, S. and Scott, C. (2000) 'Para nosotros, esta agua es vida.' El riego en condiciones adversas: Los usuarios de aguas residuales en Irapuato, México. In: Scott, C.A., Wester, P. and Marañón-Pimentel, B. (eds) *Asignación, productividad y manejo de recursos hídricos en cuencas*. IWMI, Serie Latinoamericana No. 20. IWMI, Mexico City.

Coward, E.W. Jr (1986) State and locality in Asian irrigation development: the property factor. In: Nobe, K.C. and Sampath, R.K. (eds) *Irrigation Management in Developing Countries: Current Issues and Approaches*. Studies in water policy and management no. 8. Westview Press, Boulder, Colorado, pp. 491–508.

Dinar, A., Rosegrant, M.W. and Meinzen-Dick, R. (1997) Water Allocation Mechanisms. Principles and Examples. World Bank Policy Research Working Paper 1779. World Bank and International Food Policy Research Institute, Washington, DC.

DWAF (1997) *White Paper on a National Water Policy for South Africa.* Department of Water Affairs and Forestry, Pretoria.

Edmunds, D. and Wollenberg, E. (2001) A strategic approach to multistakeholder negotiations. *Development and Change* 32, 231–253.

Falkenmark, M. and Lundqvist, J. (1998) Towards water security: political determination and human adaptation crucial. *Natural Resources Forum* 21, 37–51.

Frederiksen, H.D. and Vissia, R.J. (1998) *Considerations in Formulating the Transfer of Services in the Water Sector.* IWMI, Colombo, Sri Lanka.

Gleick, P. (1998) The human right to water. *Water Policy* 1, 487–503.

Global Water Partnership (GWP) (2000) *Towards Water Security: A Framework for Action.* GWP, Stockholm.

Green, C. and Warner, J. (2000) Flood hazard management: a systems perspective. Paper presented at the 9th Stockholm Water Symposium, SIWI, 9–12 August.

Haddad, B.M. (2000) *Rivers of Gold. Designing Markets to Allocate Water in California.* Island Press, Washington, DC.

Held, D. (1995) *Democracy and the Global Order: From the Modern State to Cosmopolitan Governance.* Polity Press, Cambridge.

Homer-Dixon, T.F. (1999) *Environment, Scarcity and Conflict.* Princeton University Press, Princeton, New Jersey.

Huffaker, R., Whittlesey, N. and Hamilton, J.R. (2000) The role of prior appropriation in allocating water resources in the 21st century. *International Journal of Water Resources Development* 16, 265–273.

Huppert, W. and Urban, K. (1998) *Analysing Service Provision. Instruments for Development Cooperation Illustrated by Examples from Irrigation.* Schriftenreihe der GTZ, No. 263. GTZ, Wiesbaden.

Huppert, W., Svendsen, M. and Vermillion, D. (2001) *Governing Maintenance Service Provision in Irrigation – A Guide to Institutionally Viable Maintenance Strategies.* GTZ, Eschborn.

Hussain, I. and Hanjra, M.A. (2004) Irrigation and poverty alleviation: review of the empirical evidence. *Irrigation and Drainage* 53, 1–15.

International Irrigation Management Institute (IIMI) (1997) *Program of Support Systems for Local Management in Irrigation.* Final Report submitted to BMZ and GTZ, November 1997. IWMI, Colombo, Sri Lanka.

IIMI (1998) *Privatization and Self Management of Irrigation, Phase II.* Final Report submitted to BMZ and GTZ, March. IWMI, Colombo.

Kijne, J.W., Barker, R. and Molden, D. (eds) (2003) Water productivity in africulture: limits and opportunities for improvement. *Comprehensive Assessment of Water Management in Agriculture Series.* No. 1. CAB International, Wallingford, UK.

Kloezen, W. (1999) Water markets between Mexican water user associations. *Water Policy* 1, 437–455.

Landry, C. (1998) Market transfers of water for environmental protection in the western United States. *Water Policy* 1, 457–469.

Lee, T.R. and Jouravlev, A.S. (1998) *Prices, Property and Markets in Water Allocation.* United Nations Economic Commission for Latin America and the Caribbean, Santiago.

Luckham, R., Goetz, A. and Kaldor, M. (2000) *Democratic Institutions and Politics in Contexts of Inequality, Poverty, and Conflict. A Conceptual Framework.* IDS, Brighton, UK.

Mehta, L. (2000) *Water for the Twenty-first Century: Challenges and Misconceptions.* IDS, Brighton, UK.

Michelsen, A.M. and Young, R.A. (1993) Optioning agricultural water rights for urban water supplies during droughts. *American Journal of Agricultural Economics* 75, 1010–1020.

Perret, S.R. (2002) Water policies and smallholding irrigation schemes in South Africa: a history and new institutional challenges. *Water Policy* 4, 283–300.

Perry, C., Rock, M. and Seckler, D. (1997) *Water as an Economic Good: A Solution, or a Problem?* IIMI Research Report 14. International Irrigation Management Institute, Colombo, Sri Lanka.

Rosegrant, M.W. and Binswanger, H.P. (1994) Markets in tradable water rights: potential for efficiency gains in developing country water resource allocation. *World Development* 22, 1613–1625.

Schreiner, B. and van Koppen, B. (2001) From bucket to basin: poverty, gender and integrated water management in South Africa. In: Abernethy, C. (ed.) *Intersectoral Management of River Basins: Proceedings of an International Workshop on 'Integrated*

Water Management in Water-stressed River Basins in Developing Countries: Strategies for Poverty Alleviation and Agricultural Growth'. IWMI and DSE, Colombo, Sri Lanka, pp. 15–69.

Scott, C.A. and Garcés-Restrepo, C. (2001) Conjunctive management of surface water and groundwater in the middle Río Lerma Basin, Mexico. In: Biswas, A.K. and Tortajada, C. (eds) *Integrated River Basin Management: The Latin American Experience.* Oxford University Press, New Delhi, pp. 176–198.

Shah, T. (1993) *Groundwater Markets and Irrigation Development: Political Economy and Practical Policy.* Oxford University Press, Bombay.

Shah, T. (2000) Water against poverty: livelihood-oriented water resource management. In: Mollinga, P.P. (ed.) *Water for Food and Rural Development: Approaches and Initiatives in South Asia.* Sage, New Delhi, pp. 38–68.

Shah, T., van Koppen, B., Merrey, D.J., de Lange, M. and Samad, M. (2002) *Institutional Alternatives in African Smallholder Irrigation: Lessons from International Experiences with Irrigation Management Transfer.* IWMI Research Report 60. IWMI, Colombo, Sri Lanka.

Simpson, L. and Ringskog, K. (1997) *Water Markets in the Americas.* The World Bank, Washington, DC.

Svendsen, M., Trava, J. and Johnson III, S.H. (2000) A synthesis of benefits and second-generation problems. In: Groenfeldt, D. and Svendsen, M. (eds) *Case Studies in Participatory Irrigation Management.* World Bank Institute Learning Resources Series. World Bank Institute, Washington, DC, pp. 139–157.

Thobani, M. (1997) Formal water markets: why, when and how to introduce tradable rights in developing countries. In: *Proceedings of Seminar on Economic Instruments for Integrated Water Resources Management: Privatization, Water Markets and Tradable Water Rights.* Inter-American Development Bank, Washington, DC.

van Koppen, B. (1998) *More Jobs per Drop. Targeting Irrigation to Poor Women and Men.* Royal Tropical Institute, Amsterdam.

van Koppen, B. (2000) *From Bucket to Basin: Managing River Basins to Alleviate Water Deprivation.* IWMI, Colombo, Sri Lanka.

van Koppen, B., Jha, N. and Merrey, D.J. (2003) (forthcoming) *Redressing Racial Inequities through Water Law in South Africa: Revisiting Old Contradictions?* Comprehensive Assessment Research Paper 3. Comprehensive Assessment Secretariat, Colombo, Sri Lanka.

Vermillion, D. and Merrey, D.J. (1998) What the 21st century will demand of water management institutions. *Journal of Applied Irrigation Science* 33, 165–187.

Wester, P. and Warner, J. (2002) River basin management reconsidered. In: Turton, A. and Henwood, R. (eds) *Hydropolitics in the Developing World: A Southern Africa Perspective.* African Water Issues Research Unit, Pretoria, pp. 61–71.

Wester, P., Merrey, D.J. and de Lange, M. (2003) Boundaries of consent: stakeholder representation in river basin management in Mexico and South Africa. *World Development* 31, 797–812.

Yoder, R. (1994) *Locally Managed Irrigation Systems: Essential Tasks and Implications for Assistance, Management Transfer and Turnover Programs.* International Irrigation Management Institute, Colombo, Sri Lanka.

Young, R.A. (1986) Why are there so few transactions among water users? *American Journal of Agricultural Economics* 68, 1143–1151.

Zwarteveen, M. (1994) Gender Issues, Water Issues: A Gender Perspective to Irrigation Management. Working Paper No. 32. IIMI, Colombo, Sri Lanka.

Zwarteveen, M. (1997) Water: from basic need to commodity. A discussion on gender and water rights in the context of irrigation. *World Development* 25, 1335–1351.

Index

Note: page numbers in *italics* refer to figures, tables and boxes.

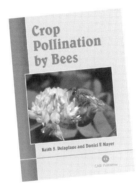
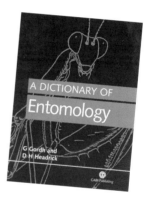
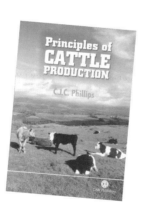